Bioregulators

ACS SYMPOSIUM SERIES 257

Bioregulators

Chemistry and Uses

Robert L. Ory, EDITOR
U.S. Department of Agriculture

Falk R. Rittig, EDITOR
BASF Aktiengesellschaft

Based on a symposium sponsored by
the Division of Agricultural and Food Chemistry
at the 186th Meeting
of the American Chemical Society,
Washington, D.C.,
August 28–September 2, 1983

American Chemical Society, Washington, D.C. 1984

Library of Congress Cataloging in Publication Data

Bioregulators, chemistry and uses.
(ACS symposium series, ISSN 0097-6156; 257)

"Based on a symposium sponsored by the Division of Agricultural and Food Chemistry at the 186th Meeting of the American Chemical Society, Washington, D.C., August 28–September 2, 1983."

Bibliography: p.
Includes indexes.

1. Plant regulators—Congresses.

I. Ory, Robert L., 1925– . II. Rittig, Falk R., 1939– . III. American Chemical Society. Division of Agricultural and Food Chemistry. IV. Series.

SB128.B56 1984 631.8 84-10987
ISBN 0-8412-0853-0

PRINTED IN THE UNITED STATES OF AMERICA

ACS Symposium Series

M. Joan Comstock, *Series Editor*

FOREWORD

The ACS Symposium Series was founded in 1974 to provide a medium for publishing symposia quickly in book form. The format of the Series parallels that of the continuing Advances in Chemistry Series except that in order to save time the papers are not typeset but are reproduced as they are submitted by the authors in camera-ready form. Papers are reviewed under the supervision of the Editors with the assistance of the Series Advisory Board and are selected to maintain the integrity of the symposia; however, verbatim reproductions of previously published papers are not accepted. Both reviews and reports of research are acceptable since symposia may embrace both types of presentation.

CONTENTS

PREFACE

WORLD POPULATION IS EXPECTED to almost double by the year 2020, and cause the amount of land used for housing to increase while the amounts available for growing the food and fiber needed for these additional people decreases. Innovations in agriculture will be needed to meet the challenge. Not only will vast increases in production of basic crops be necessary, but losses incurred during postharvest handling and processing will have to be reduced or eliminated to ensure that greater amounts of these crops reach the consumer. Improvements in plant breeding, tissue culture, and genetic engineering; breakthroughs in plant and animal growth regulators; and other new technologies will have to be developed.

Agriculture's bioregulators might be compared to medicine's "miracle wonder drugs." They are used in very small quantities—perhaps once a year or once during the plant's lifetime—to induce increases in yield, shelf life, and quality; to improve resistance to stress and microbial or insect damage; or to eliminate weeds by altering and controlling physiological processes in the plants.

This 21-chapter volume contains up-to-date reports on the chemical aspects of synthetic bioregulators; on applications to improve the major cereals, oilseeds, legumes, citrus, sugar crops, tree fruits, nuts, and nonedible crops such as cotton and guayule (a promising new source of latex for rubber production); on physiological factors such as water relations, temperature extremes, and environmental stress; and on improved resistance to insects, fungi, and microbial toxins. Several chapters covering research on natural bioregulators and applications in tissue culture studies provide a balanced coverage of this subject.

Much remains to be learned about bioregulators, and we realize that this book does not cover everything. This field is opening new areas of research for industrial chemists involved in the synthesis of new analogs, for plant physiologists, molecular biologists, and biochemists studying their action, for agronomists searching for improved crops, and for those involved with pesticides, herbicides, and the various steps during the postharvest handling and processing of agricultural crops. We hope that this volume will kindle new ideas and stimulate new and greater research on applications of bioregulators in agriculture.

ROBERT L. ORY
U.S. Department of Agriculture
New Orleans, Louisiana

FALK R. RITTIG
BASF Aktiengesellschaft
Limburgerhof, Federal Republic
of Germany

March 1, 1984

1

Introduction

ROBERT L. ORY

Southern Regional Research Center, Agricultural Research Service, U.S. Department of Agriculture, New Orleans, LA 70179

FALK R. RITTIG

BASF Aktiengesellschaft, Landwirtschaftliche Versuchsstation, D-6703 Limburgerhof, Federal Republic of Germany

> "Can you do addition?" the White Queen asked. "What's one and one and one and one and one and one and one?" "I don't know said Alice, "I lost count." "Can <u>you</u> do sums?" Alice said, turning suddenly on the White Queen... The Queen gasped and shut her eyes. "I can do addition", she said, "if you give me time--"
>
> Lewis Carroll, in
> Through the Looking Glass

Lewis Carroll's statement almost describes the challenge facing agricultural scientists and engineers in the next 3-4 decades. World population is expected to double. There is a finite amount of land available on earth. Thus, as land now under cultivation is converted into real estate, the farmers who produce our food and fiber will have to develop new unique methods to get more food from less land, and those involved with the post-harvest handling/processing of these crops will have to find better ways to eliminate waste and losses due to damage, disease, spoilage, and invasion by various pests during the post-harvest chain of events. In other words, they must add much more to the food supply, with less land, less energy, and before time runs out.

The large increases in yields of American agricultural production over the past century, with only 3% of the labor force engaged in farming (compared to 35% in 1910), and Dr. Norman Borlaug's "Green Revolution" that virtually changed Mexico from an importer of wheat to an exporter within a generation, have been acclaimed all over the world. These increases were due largely to improved breeding of crops, mechanization of farming/harvesting methods, and increased use of agrochemicals (fertilizers, pesticides and herbicides). Research in these areas must continue but they can achieve just so much with the amounts of available farm land decreasing. Production of food and fiber must keep ahead of population growth. New methods for increasing crop yields in a single season must be developed and new more efficient methods for decreasing post-harvest losses must be found. This means developing new biotechnologies to

0097-6156/84/0257-0001$06.00/0

alter physiological and biochemical processes in crops and plants, inhibit or reduce activity of enzymes that produce undesirable effects in stored crops and/or increase activity of those that are desirable, increase the resistance of plants to stress conditions; i.e.: excessive heat, frost, high winds, salinity in water, damages caused by mechanical harvesting, insects, mycotoxin invasion, and other causes of food and fiber losses. Some countries experiencing food shortages today can produce or import sufficient quantities, but post-harvest handling and distribution losses reduce this amount by almost half. Improved handling and quality maintenance is, therefore, an important goal for increasing available food and fiber but this will not be enough.

Fertilizers also play a major role in the high yields produced by U.S. agriculture today but these are expected to change significantly in the future. The Fertilizer Institute estimates that by the 1990's, natural gas will no longer be a source of nitrogen fertilizers (1). The sources and kinds of fertilizers will have to change but, again, increased amounts of fertilizers alone cannot produce the quantities of food and fiber that will be needed in the 21st century.

Bioregulators (plant growth regulators) are hormone-like chemicals that can increase yields with one application, alter growth patterns, nutritional components, and resistance to different kinds of stress (cold, heat, drought, insects, disease) when applied at extremely low levels. They can be as beneficial or economically important as herbicides, pesticides and fertilizers and, because they are used in such small doses, they generally have less impact on the environment than most pesticides. The major benefit of these chemicals is that the plant's or crop's physiological functions are under control of the farmer. Nickell (2) defines plant growth regulators as "either natural or synthetic compounds that are applied directly to a target plant to alter its life processes or its structure to improve quality, increase yields, or facilitate harvesting." The term "plant hormone" refers more appropriately to naturally occurring plant growth substances such as gibberellins, auxins, cytokinins, abscisic acid, growth inhibitors and ethylene (2). The metabolism, modes of action, and physiological functions of these compounds were recently reviewed by Wareing (3) in 60 papers presented at the International Plant Growth Substances Association Conference held in Wales in July 1982.

The term "regulator" is quite broad and includes both synthetic and natural substances. Both are covered in the following chapters but the greatest challenge facing the agrochemical industry is to continue to synthesize newer, more efficient compounds than are available today. Also, unlike the auxins, gibberellins and cytokinins which are present and function in virtually all plants, the synthetic bioregulators may be specific for certain crops and for specific physiological functions. They are not as old as the natural plant bioregulators but their future importance is unquestioned. These substances are being acknowledged as a primary means of providing the food and fiber needed in the 21st century and beyond.

In 1975, plant growth regulators made up less than 5% of world agrochemical sales (herbicides, insecticides and fungicides) but at least 29 U.S. companies and several in Europe were reportedly developing substances with chemical and biological activity (4). Herbicides are unlikely to have to cope with totally new crops, although they will be required for control of new weeds that will surely emerge in the future (5). Herbicides represent the extreme in plant growth regulators whereas both natural and synthetic growth regulators interefere with the plant's endogenous hormonal pattern (5). The effects can be morphogenetic or metabolic so that the potential is extremely diverse. All physiological processes, plus crop yields, can be affected. In the next 10-25 years, Marvel (6) predicts that plants will be induced to be resistant to pesticides, immune to disease, tolerant to heat, cold, and brackish water, and to produce more protein.

Searching for Bioregulators

The concept of applying chemicals to regulate plant growth is not new. In fact, the use of plant growth regulators for controlling various physiological processes in plants has been known for more than five decades. Only in more recent times, however, has the regulation of characteristics different from growth or morphology of plants led to the use of the term "bioregulators" for these chemicals (7), applying this term to both natural and synthetic compounds (8).

Whereas for a broader application only synthetic bioregulators will play a major role in agriculture, natural phytohormones cannot be excluded when discussing bioregulation (9). These chemicals are only partially in use today but they will hold their niche in the future. The synthetic compounds are structural analogues of nature's phytohormones. The system of classifying growth regulating compounds into auxins, gibberellins, cytokinins, abscisins and ethylene generators also applies to synthetic chemicals. Close chemical relationship (structural analog, derivative, generator) and comparable biological activity (agonist, antagonist) make possible a further classification.

Because of the relatively simple way other crop protection agents can be developed and applied, often with almost immediate effects, plant bioregulators have always been overlooked as of lesser importance, especially when compared to herbicides. Within the last few years, however, agriculture has turned its interest increasingly to plant bioregulators because of interest stimulated by the successful introduction of bioregulator use on some important crops. It is anticipated that the use of bioregulation to increase yields and improve quality and safety of plant products will grow. Bioregulation was the subject of the first of a series of long-range planning seminars of the U.S. Department of Agriculture, Agricultural Research Service, in April, 1982, at Bethesda, Maryland.

There is a difference of opinion on the economic importance of plant bioregulators. It is clear that their present use is centered in the highly industrialized countries, such as the USA or Western Europe. Application is also limited to experienced agricultural management as an ultimate tool when all other measures for increasing yield (soil tillage, irrigation, fertilization, crop protection) have been successfully applied as prerequisites. Finally, the crops on which they are now used are grown on an industrial basis, such as sugar cane, pineapples, hevea and tobacco. Despite these prior conditions and the availability of mostly qualitative testing systems, there is great interest in plant bioregulators from the scientific community. Research into understanding the mechanisms of endogenous plant hormone action and the potential for crop manipulation by plant bioregulators has been active, and seems to be increasing. Plant bioregulators have the smallest market among all the major pesticide categories, but they also show the fastest growth (19). Considerable effort is being devoted to the field of industrial research by the majority of the producers of crop protection agents in the USA, Western Europe and Japan, plus some in the major countries of Eastern Europe.

Despite years of research by both private enterprise and government agencies, only a few products play a major role in this market, amounting to about 5% of the total market of crop production chemicals. Recommended crop uses for plant bioregulators are suprisingly few and the number of chemicals on the market is still comparatively small (10). Products that have been accepted by the market are ripeners in sugar cane, stimulants of latex flow in Hevea, and antilodging compounds in cereals. Most recently a plant bioregulator in cotton was developed for limiting undesired vegetative growth and increasing yield in various countries (11), and in Taiwan, a plant bioregulator is in use for programming rice seedling propagation (12).

The only major product which is of less importance in the USA, but which has been used for about two decades in Europe (including the USSR) on more than half of the wheat acreage, is chlormequat (13). This chemical prevents lodging, allows increased fertilizing and maximizes small grain yields. As the first example of a successfully applied and widely used synthetic plant growth regulator, chlorcholine chloride has encouraged the search for other synthetic gibberellic acid antagonists based on the common feature of an onium ion; i.e., a charged central atom. In the beginning, structure activity relationship studies on a more qualitative basis dealt with just the trimethyl ammonium moiety, varying the fourth alkyl group attached to the central atom carrying the positive charge, because Tolbert (14) postulated the trimethyl ammonium structure to be essential for growth regulating activity. Later, a number of bioregulators were developed which do not follow Tolbert's hypothesis; e.g., the hydrazinium derivative CMH (15). CCC was an ideal compound to investigate structure activity relationships since the synthesizing chemist could vary the charged central atom, substituting nitrogen with sulfur or phosphorus (11). In some cases two methyl groups were used to close a five- or six-membered ring that also contained other hetero atoms, thus forming a morpholinium salt (16). Chapter 2 describes these synthetic aspects in detail, plus the application of theoretical approaches like Rekker's Fragmental Constants or the First Order Molecular Connectivity Index. For the chemist it is somewhat consoling that the thousands of chemicals synthesized and not found to be more active than existing bioregulators, may be used to retrospectively serve as input for theoretical QSAR considerations. For such a synthesis, the chemist can sometimes look at microbial metabolites that possess biological activity as a lead. Using products from fungi as templates for tomorrow's bioregulators is described in Chapter 14.

With the synthesis of a potentially active new compound, however, the difficulties do not end. The screening for and development of plant bioregulators is still regarded as more expensive than the testing of a fungicide or herbicide (17). Difficulties arising from the necessarily time-controlled application of bioregulators to influence certain characteristics in the development of cereals are demonstrated in Chapter 4. Another potential disadvantage is the fact that an impact of any kind on a biological system can result in a change not only of the desired property (i.e., yield increase) but it also might affect resistance, quality, taste, maturity, or other parameters. Of course, breeding a new line will be the ultimate solution to improve a crop, but breeding is a time consuming process; it may take ten years or more to establish a new variety. Meantime, bioregulators can be applied to achieve the desired goal or they may be overimposed during crop breeding to produce an immediate effect.

Physiological ingenuity and analytical expertise, as described in Chapters 3 and 12, are necessary to come to statistically satisfying results. In addition to the usual variability of biological systems, very often large differences in the reactions of different plant species are to be expected. Environmental factors have significant influence on the activity of plant bioregulators. Thus, in contrast to crop protectants developed to kill a target organism (pests, diseases, weeds), a greater significance is necessary in the case of plant bioregulators. This is documented by more expensive and delicate test systems for this class of compounds. Here the use of near infrared reflectance emission spectrophotometry (Chapter 18) can serve as a means not only for screening bioregulators, but also to set up an automatic working test ideally suited for the industry and for applying computer processing.

As compounds from the latest generation of bioregulators are active at ppm levels, proper application is an essential condition for their success. One possibility for avoiding problems involved with a homogenous spray cover is to bring bioregulators directly to their site of action. Best means for this would be a seed treatment (10), providing the active chemical without producing unnecessary problems for the farmer. Formulation specialists are working in this field on controlled release formulations. Chapter 9 illustrates that the use of a drench application, to a great extent, can avoid environmental problems connected with persistence of a bioregulator in the environment. Such persistence would have been detrimental to the further development of a specific bioregulator that is important for reducing undesired vegetative growth in apple trees. An even better solution was the incorporation of this active ingredient into a banding laid around the trunk of the tree. By this way the ingredient is slowly released at a controlled rate without any undesired environmental effects (18).

Applications in Agricultural Crops

The large expected increase in world population will probably place greatest demands on the cereal grains. These are the primary staples; they are low cost sources of food calories and, for many nations, also a major source of protein. Additional sources of low cost edible protein will have to come from legumes, pulses, and oilseeds. Research on photosynthesis has broadened our understanding of this process and improvements in photosynthesis will play a significant role in increasing crop yields. We now understand more about photosynthesis than we do about various aspects of storage of crops. What happens during post-harvest and new methods for overcoming post-harvest losses are areas requiring more effort. Other agrochemicals (pesticides, herbicides, fungicides, pheromones) will still be necessary but so will the plant growth substances, not only to increase yields but also to alter the biochemical and physiological characteristics in such a way as to render the crops more resistant to the various stresses of post-harvest handling and storage.

Other chapters will describe our current knowledge of both synthetic and natural plant bioregulators, their applications for both food and fiber crops and on increasing resistance to insects and stress factors. The chemical aspects of a possible relationship between chemical structural features and growth regulating activities are covered in Chapter 2. This is followed by a detailed comparison of results from laboratory and field evaluation tests with bioregulators.

Since food is a constant necessity and will continue to be as world population grows, applications to improve yields and/or quality of edible crops are described in the next ten chapters. Cereal grains are covered in 4 and 5; oilseeds and legumes in 6-8; sugar crops, fruits and nuts in Chapters 9-12. Uses of bioregulators to improve resistance to fungi and insects are then described in Chapters 13-14. Applications for non-edible crops, such as cotton and latex, and for induction of resistance to various stress factors are covered in Chapters 17-20. Natural regulating substances and phytoalexins (Chapter 21) and allelochemicals (Chapter 15) complete the balanced coverage of this very broad and growing subject. For more detailed information on any of these subjects, readers may consult references cited by the authors in their respective manuscripts.

The development of plant bioregulators in the future will continue to be an even greater challenging task for the experimental biologist who will, in a concerted action, depend upon the assistance of and cooperation with chemists, physicochemists, and plant physiologists. Industry has reached maximum yields as far as our present technology is concerned. To make another quantum jump we must use tomorrow's tools - plant bioregulators (20).

Literature Cited

1. Lepkowski, W. Chem. and Eng. News 1982, 60 (no. 47), 8-16.
2. Nickell, L. P. "Plant Growth Regulators. Agricultural Uses"; Springer-Verlag: Berlin-Heidelberg-New York, 1982; p. 1.
3. Wareing, P. F. "Plant Growth Substances 1982"; Academic Press: New York.
4. Anon. Farm Chemicals 1975, 138 (no. 9), 45-48.
5. Krieger, J. H. Chem. and Eng. News 1982, 60 (no. 51), 9-23.
6. Marvel, J. T. Middle Atlantic Regional Meeting, American Chemical Society 1983. White Haven, PA, April 6-8, Abs. 6.
7. Poling, S. M.; Hsu, W.-J.; Yokoyama, H. Phytochemistry 1976, 15, 1685-7.

8. Jung, J. Arzneim.-Forsch./Drug Res. 1980, 30 (11), 11a, 2.

9. Draber, W., in "Chemie der Pflanzenschutz- und Schädlingsbekämpfungsmittel"; Wegler, R., Ed.; Springer-Verlag Berlin-Heidelberg-New York, 1977; Vol. IV, pp. 1-63.

10. Oplinger, E. S. Proc. Beltsville Symposium VIII: Agricultural Chemicals of the Future, 1983, in press.

11. Zeeh, B.; König, K. H.; Jung, J. Kemia-Kemi 1974, 1, 621-3.

12. Jung, J.; Koch, H.; Rieber, N.; Würzer, B. J. Agronomy & Crop Science 1980, 149, 128-136.

13. Jung, J.; Schott, P. E. Z. f. Pflanzenkrankheiten und Pflanzenschutz, Sonderheft VII, 1975, 333.

14. Tolbert, N. E. J. Biol. Chem. 1960, 235, 475.

15. Jung, J. Zeitschr. Acker- und Pflanzenbau, 1973, 137, 258.

16. Jung, J. Zeitschr. Acker- und Pflanzenbau, 1970, 131, 325.

17. Sacher, R. M., in "Chemical Manipulation of Crop Growth and Development"; McLauren, J. S., Ed.; Butterworths, London, 1982, pp. 13-15.

18. Hield, H. Proc. 10th Annual Meeting Plant Growth Regulator Society of America, East Lansing, Mich. USA, 1983, in press.

19. Anon. C&EN, 1980, 58 (no. 17), 13.

20. Anon. Farm Chemicals 1983, 146 (no. 9), 41.

RECEIVED April 4, 1984

2

Chemical Aspects of Some Bioregulators

HUBERT SAUTER

BASF Aktiengesellschaft, Hauptlaboratorium, D-6700 Ludwigshafen, Federal Republic of Germany

The discovery of several new compounds with plant growth retarding activity was the result of many years of research at BASF on structural modification of the lead CCC. Besides mepiquat chloride, a new bioregulator for commercial use in cotton, other new ammonium-, hydrazinium- and sulfonium-compounds have been found to be very active. Structural requirements in this group of compounds showing generally high activity in wheat and cotton are reviewed. Quite different structural prerequisites are necessary for ammonium compounds to be active in rice. QSAR-studies on some recently discovered pyridinium-salts show a clear correlation between rice retarding properties and the lipophilic parameters of the substituents. Another group of ammonium salts shows activity in soybeans. In contrast to the ammonium-type compounds, which individually exhibit high growth retarding activity only in a limited number of crops, the new BASF bioregulator BAS 106 W, which belongs to the class of norbornanodiazetines, shows excellent growth regulating properties on a great variety of plant species. In this connection, the question of whether there is a common structural feature responsible for the growth regulating activities in the classes of norbornanodiazetines, triazoles and pyrimidines, is discussed.

This chapter reviews research into bioregulator design and synthesis which has been carried out at BASF for more than 15 years and is still being actively pursued. A few examples are presented of different types of chemical compounds and their structures in relation to bioregulating activity or, as is frequently the case, inactivity. This will give some idea of how synthesis chemists approach the problem of finding bioregulators. These are sometimes found by chance and sometimes more systematically by means of

0097-6156/84/0257-0009$06.00/0

molecular modification and optimization of existing lead structures.

Onium Compounds

The lead CCC. Figure 1 shows one of the more important lead structures, i.e. CCC. Tolbert has described some characteristic bioregulating effects obtained with CCC-treated wheat, particularly the development of stockier plants with shorter and thicker stems (1). This influence on plant growth and development was later found to be of great practical value (2). Figure 1 shows two further structures whose activities are just as high. However, Tolbert showed that further structural variations, even when these were minor ones, such as lengthening only one methyl group to an ethyl group, lead to a very large decrease in activity. These observations led him to the hypothesis that the trimethylammonium structure is essential for maximum activity.

Structure modification of CCC. At this stage in the developments, work began at BASF in the early 'sixties with the aim of modifying the lead structure of CCC. The aim was to try to find novel compounds which had similar, if possible even more powerful activities, a greater range of action or a different activity spectrum. It should be remembered that while CCC had good growth inhibiting effect on wheat, the corresponding effect on oats, rye and barley was substantially less pronounced. In the course of this work, well over 1 000 different compounds with a cationic structure - the so-called onium compounds - were synthesized and tested.

It soon became apparent that if the 2-chloroethyl side chain is left unchanged, the trimethylammonium group can be replaced by cationic centers with other hetero atoms for example, by the dimethylsulfonium group or by the trimethylphosphonium group (3). Although in these cases the activity is reduced as a result, it does not in principle vanish. In contrast, incorporation of an uncharged carbon atom results in a change in the molecular properties which is so pronounced that the activity is completely lost.

Also certain related hydrazinium compounds like CMH and AMH (Figure 2) possess just as high an activity as CCC (4, 5). In this case, a methyl radical of the trimethylammonium group has been replaced by the amino group NH_2.

From the point of view of molecular structures, the jump from CCC to heterocyclic onium compounds of the mepiquat-chloride type (3) seems substantially larger in comparison (Figure 2). This compound and some related structures surprisingly show even higher bioregulating activity than CCC. Particularly good results have been obtained with cotton and barley. Mepiquat-chloride is used commercially in the cotton bioregulator Pix. Together with the ethylene-releasing ethephon, it is also one of the active ingredients of the commercial growth regulator Terpal, which is used to prevent lodging in barley.

Structure-activity relationships of mepiquat type compounds. Structure-activity relationships in this group of compounds have already been reviewed by Zeeh et al. (3). The main aspects of these and introduction of a few new ones are briefly summarized. It has been shown that any substitution on the ring of the dimethylpiperidinium cation results in a loss of activity. Derivatives with two ring substituents are generally inactive in wheat.

Figure 3 shows how the activity is affected by ring size and exocyclic substituents on the nitrogen atom. In the case of the 6-membered rings, the activity decreases sharply when the methyl group is lengthened by only one carbon atom; the diethylpiperidinium salt and the methylpropylpiperidinium salt are virtually completely inactive.

The 5-membered compound with two methyl groups is somewhat less active than the corresponding 6-membered ring. However, in the series of compounds possessing 5-membered rings, the activity decreases substantially more slowly when the methyl groups in the N-position are lengthened; even the diethyl compound still shows some activity.

In the case of the compounds possessing 7-membered rings, even the dimethyl compound is virtually completely inactive.

Some 6-membered cyclic sulfonium compounds (3) also show an unusually high level of activity (Figure 4). Also a high level of activity is often retained when carbon atoms of the mepiquat cation are exchanged for additional hetero atoms, such as sulfur, oxygen or nitrogen (6) (Figure 4). All these compounds are very similar to the dimethylpiperidinium cation in respect to molecular structure and molecular size.

After all these systematic molecular variations a more detailed study of the mepiquat cation was made in order to find out at which points of the molecule and in which manner the cation could interact with its hypothetical receptor. Such a study should then help to develop novel and reasonable hypotheses and ideas for structural variations.

In an analysis of this type, the hydrophobic fragmental constants calculated by R. F. Rekker (7) were used to characterize individual atoms and atom groups within a molecule in terms of their contributions to lipophilicity or hydrophilicity. Figure 5 shows the result of this mapping for the mepiquat cation. The negative values around the cationic center indicate areas in which hydrophilic interactions are to be expected, while the CH_2 groups, which have positive values, tend rather to lipophilic interactions. Seen overall, the molecule can be clearly divided into a hydrophilic part and a lipophilic part.

On the basis of this crude model, this raised a question of whether the mepiquat cation was varied sufficiently? For the lipophilic area, any enlargement in this part of the molecule leads to a decrease in activity. But what about the hydrophilic area? In this case, going from methyl groups to longer alkyl groups likewise results in a loss of activity. However, longer alkyl groups are also more lipophilic groups, which may disturb the hydrophilic interactions in this region. On the basis of these considerations, a few compounds were synthesized in which a hydrophilic methyl group was replaced by larger, but at the same time, more hydrophilic substituents. Figure 6 shows a few examples (8). However, neither these compounds nor any similar ones showed any significant activity.

Therefore, it can be concluded that these onium compounds are subject to narrow steric restrictions in the hydrophilic area as in

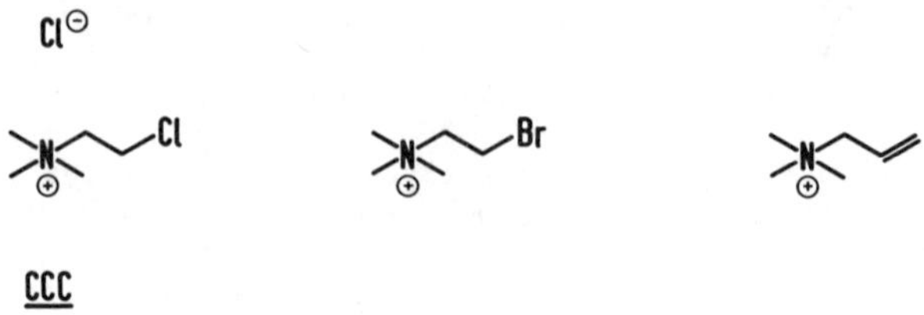

Figure 1. Trimethylammonium compounds with high growth retarding activity (1).

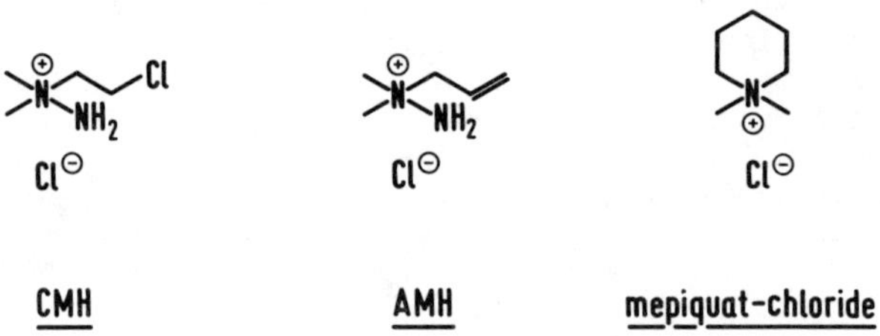

Figure 2. Modifications of CCC.

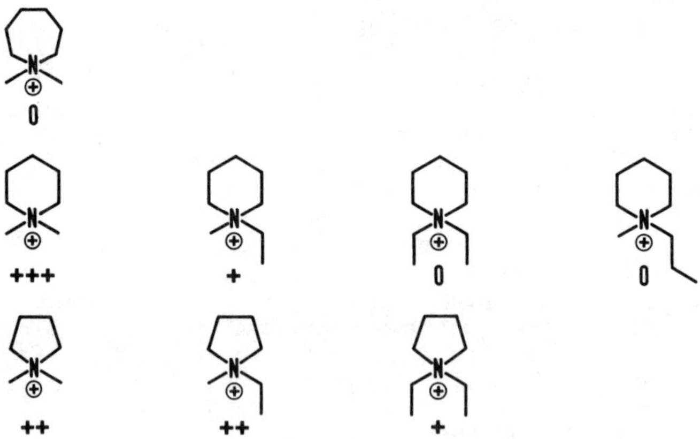

Figure 3. Structure-activity relationships of mepiquat-type compounds.

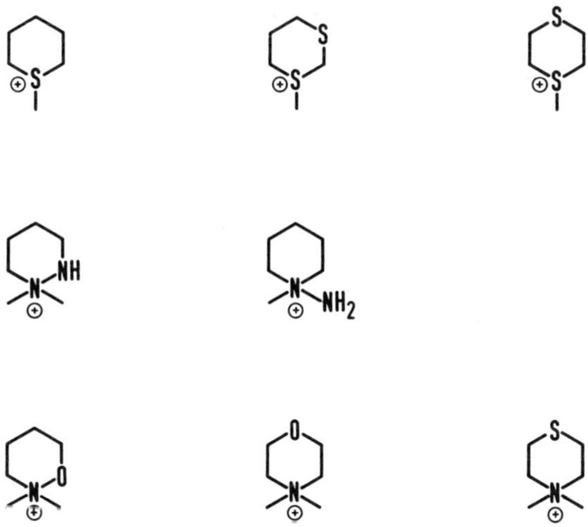

Figure 4. Highly active cyclic onium compounds (3, 6).

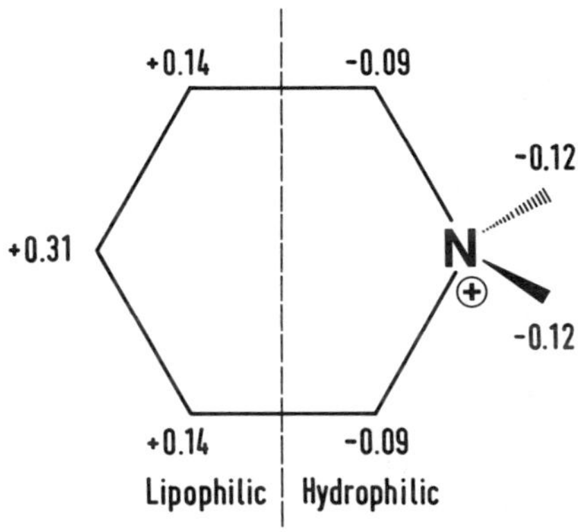

Figure 5. Lipophilic and hydrophilic regions of the mepiquat cation as given by Rekker's fragmental constants.

the lipophilic area, and high activity cannot be achieved without observing these restrictions.

To summarize the structural prerequisites which obviously have to be satisfied in order to obtain bioregulating activity of the CCC or mepiquat types: first, the molecule must contain a quaternary ammonium or phosphonium atom or a tertiary sulfonium atom. Second, the molecule must not exceed a certain size.

In the case of the ammonium compounds, which do not contain further hetero atoms, the size required can be defined in terms of the number of carbon atoms or hydrogen atoms. All C, H, N compounds which have moderate to high activity possess 6 to 8 carbon atoms and 16 to not more than 18 hydrogen atoms. So we have structure-activity relationships just by counting atoms.

The steric prerequisites for activity can also be described in somewhat more sophisticated terms by means of a topological descriptor i.e. the first-order molecular connectivity index $^1\chi$ (9). By putting together these first order connectivity indices $^1\chi$ and the corresponding activity data of more than 100 ammonium compounds, we arrived at the conclusion illustrated in Figure 7 by means of a few examples (8). All compounds showing moderate to high activity have $^1\chi$ values between 2.9 and 4.5. Within this range, only a few compounds have been found which have either very low activity or no activity at all. In contrast, compounds with $^1\chi$ values below 2.9 and above 4.5 have very little activity or are completely inactive.

All these CCC-like and mepiquat-like onium compounds exhibit a more or less similar activity spectrum. Their dwarfing activity is generally very pronounced in wheat and cotton but less pronounced in rye and barley. They are virtually completely inactive in the cases of rice and soybean. Thus, within this structurally defined group, we saw no prospect of extending the activity spectrum to include rice and soybean by varying the structure. On the other hand bioregulators for these two large scale crops are of particular practical importance and other chemical types had to be sought.

Pyridinium salts for rice. During synthetic studies involving some synthesis intermediates, it happened that a pyridinium salt which showed some growth retardation in rice was found. Its structure is shown in Figure 8 (X = 2-CH_3, 4-Cl). The first-order molecular connectivity index of the cation is 40 and hence well outside the range for the mepiquat type compounds. Accordingly, the compound is completely inactive in the case of wheat. In the rice seedling assay (10), the compound shows a KI_{50} value of 1,7 x 10^{-4} mole/l. KI_{50} is the concentration necessary for 50 % growth inhibition. Comparison with the KI_{50} value determined for mepiquat in the rice seedling assay shows that the activity of the pyridinium salt is 40 times as high.

At this stage, it appeared to be a promising approach to start from this new lead structure and optimize the rice retarding properties by systematic structure variation. As a first step, we modified the substituents of the aromatic moiety while keeping the remaining structure constant. A small series of 18 compounds was chosen.

After testing these compounds, the results were subjected to a Hansch-type multiple regression analysis (11). This showed that the

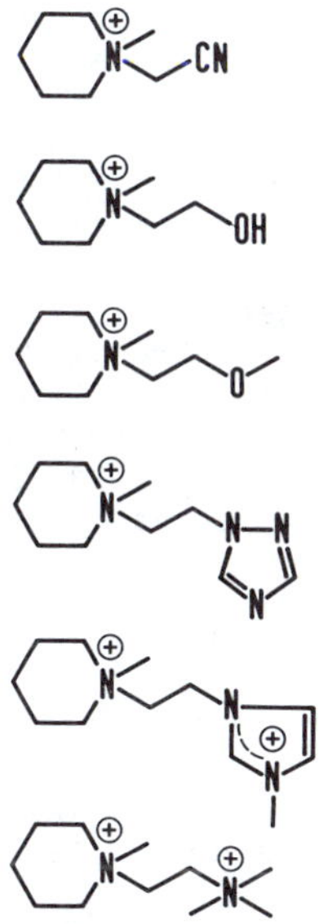

Figure 6. Replacement of one methyl group of the mepiquat cation.

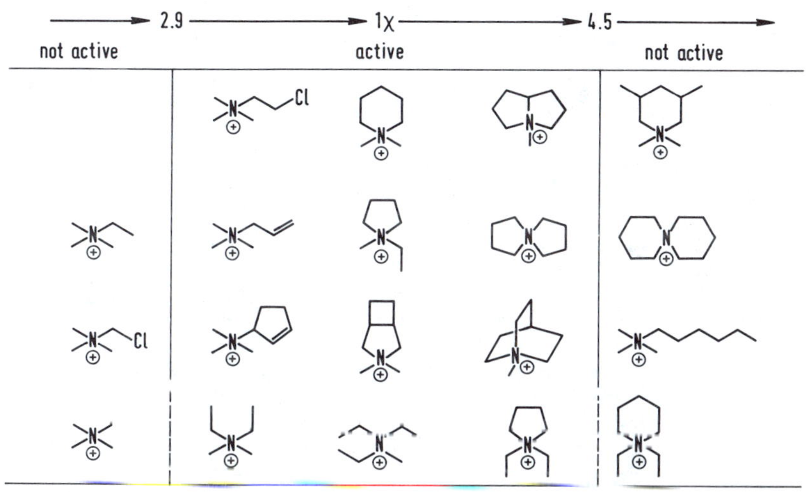

Figure 7. Molecular connectivity of ammonium compounds and growth retarding activity (wheat).

electronic and steric properties of the substituents in the aromatic moiety have no significant effect on the action. On the other hand, the lipophilicity of the substituents has a decisive effect. In Figure 8 the activity is plotted as the pI_{50} value; this is the negative logarithm to the base 10 of the KI_{50} value. A large pI_{50} value indicates high activity. The sums of the calculated (12) lipophilicity constants of substituents X $\Sigma\pi_X$ are plotted along the x axis. Starting from the relatively hydrophilic 3-acetylamino derivative, the activity initially increases with increasing lipophilicity, and then reaches a plateau. The dependence of the activity on the lipophilicity can be readily described as a function of $\Sigma\pi_X$ and $(\Sigma\pi_X)^2$ of the substituents, by means of the equation given in Figure 8, where the numbers in parentheses are the 95 % confidence intervals. The correlation coefficient is r = 0.94. The structure-activity relationships can also be described by a bilinear equation:

$$pI_{50} = 0.58\ (\pm\ 0.15)\Sigma\pi_X\ -\ 0.49\ (\pm\ 0.21)\ \lg\ (\beta \cdot 10^{\Sigma\pi_X} + 1) + 2.86\ (\pm\ 0.12)$$

$\lg \beta = -\ 1.31 \quad n = 18 \quad s = 0.15 \quad r = 0.95 \quad F = 47.8$

In this case, as well as in many similar ones (13, 14), the data seem to fit slightly better than in the parabolic equation of Figure 8.

However, both results unfortunately mean that there appears to be no possibility of improving the rice-dwarfing activity of this type of compounds substantially beyond that of the initial lead structure by varying the substituents X.

Furthermore, no substantial increase in activity was achieved when the chain between the aromatic moiety and the pyridinium moiety was modified or when substituents were introduced into the pyridine ring. Therefore, all attempts to optimize our pyridinium lead structure came to a dead end and did not produce derivatives which were sufficiently active and could be used in practice as rice retarders.

Ammonium compounds for soybeans. In contrast, work involving replacement of the pyridine ring by other heterocyclic amines was more successful. We introduced a part of the mepiquat molecule in the form of the piperidinium unit. Figure 9 shows some of the structures which resulted.

This new type of bioregulating ammonium compounds (15) does not show any mepiquat-like activity in wheat and cotton. In the case of rice too, their activities do not surpass those of the pyridinium salts already discussed. However, they surprisingly show very powerful growth inhibition in soybean after application to leaves. The treated plants are shorter and stockier and have an intense green color. In this series, the soybean activity increases from left to right. In contrast to the structure-activity relationships in the mepiquat series, in this type of compounds, lengthening the methyl group leads initially to an increase in activity.It should be noted that within this large group of cationic compounds there are individual sub-groups which show high activity on individual plant species only depending on the individual structure types.

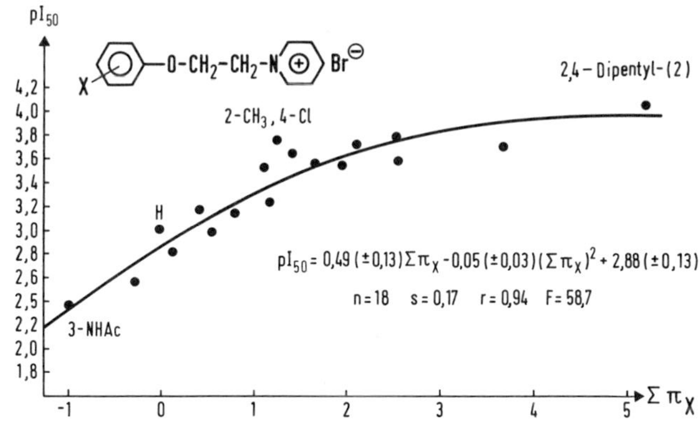

Figure 8. Quantitative structure-activity relationships: Aryloxyalkyl pyridinium bromides/rice seedling assay (11).

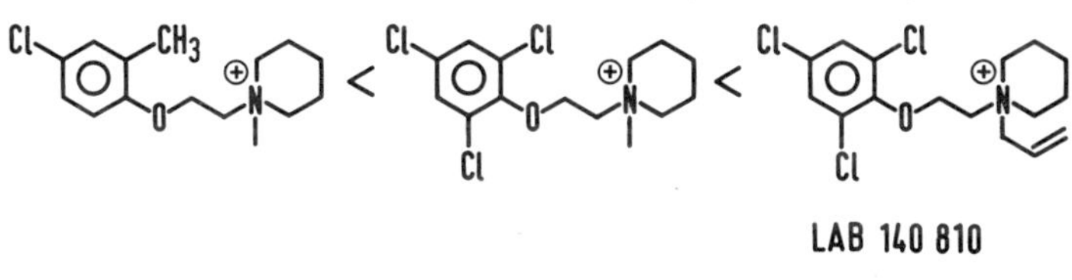

Figure 9. Ammonium compounds with soybean growth retarding activity (15).

Nitrogen Heterocycles

Nitrogen heterocycles are a completely different group of highly active bioregulators which generally exhibit broad activity spectra for a large number of plant species. From the point of view of the chemical structure, this group comprises three types of nitrogen heterocycles, whose central features are shown in Figure 10. The first type is characterized by the 4-membered diazetine ring, the essential feature of the second type is a 5-membered, 1,2,4-triazole ring, and the third type contains the 6-membered pyrimidine ring.

Norbornanodiazetines. The structure type of norbornanodiazetines is a speciality of BASF (16). Figure 10 shows the structure of the most prominent member; namely, BAS 106 W. This bioregulator with the common name tetcyclacis is currently being developed for a seed-soaking treatment for rice (17). In biological tests, this compound was found to be a highly active bioregulator with a very broad activity spectrum (18).

Variation of this novel lead structure led to a number of similar compounds, some of which also had high activity. Structure-activity studies (19) revealed that apparently, the presence of the 5-membered triazoline ring is not essential to activity. It can be replaced by other quite different substituents. On the other hand it should be emphazised that the diazetine ring seems to be quite essential for high activity.

Mode of action. Investigations into the mode of action of tetcyclacis (20, 21) show that its bioregulating action is most probably based on the inhibition of the biosynthesis of gibberellins. In the cell-free enzyme system from pumpkin endosperm, it had been possible to show that tetcyclacis, at a molar concentration as low a 10^{-7}, specifically inhibits the oxidation steps from kaurene via kaurenol to kaurenal and kaurenoic acid (20, 21).

Interestingly, it was also possible to show that a whole series of different triazole compounds block the same steps with as high a specificity, in 10^{-7} molar solution (20, 21). One example of this triazole type compounds is the ß-triazolyl-ketone LAB 117 682 (22), which, on many plant species, shows bioregulating activities which are qualitatively similar to those observed with nobornanodiazetines.

A few years previously, it had been shown that the well-known bioregulator ancymidol inhibits exactly the same biosynthetic steps (23). This led to the pyrimidines, the third chemical class having the same mode of action or, more exactly, the same points of inhibition in gibberellin biosynthesis.

The common structural feature. From the chemical point of view, the question now arises as to whether these common points of inhibition in all three structure types are accidental, or whether the norbornanodiazetines, the triazoles and the pyrimidines possess common structural features which may give rise to a similar interaction with a common point on the target receptors.

These three chemical classes do indeed possess a common structural feature, i.e. a lone electron pair on the sp^2-hybridized nitrogen atom of the particular heterocyclic ring. It should be emphasized that this electron pair is located in each case at the periphery of the molecule and is hence easily accessible sterically from outside. Lone electron pairs of this special type are just the kind which act as ligands and form complexes with heavy metal cations. One may ask, however, what this has to do with the inhibition of the oxidation of kaurene.

Coolbaugh et al. (24) were able to show that the enzymes required for the oxidation of kaurene to kaurenoic acid belong to the mixed function oxygenases of the cyctochrome P-450 type. Their active site contains a protohaem iron, which normally binds oxygen as a sixth ligand. They found some evidence that ancymidol displaces the oxygen from the iron and takes its place, thus competing with the natural substrate oxygen (cf. Figure 11). Such a ligand exchange, in which oxygen is replaced by a nitrogen heterocycle, gives characteristic UV difference spectra, the so-called type II spectra. A spectrum of this type has also been observed in the case of ancymidol.

Similarly, the triazole fungicide dichlorobutrazol gives type II difference spectra with cytochrome P-450 preparations obtained from rat liver microsomes, as recently reported (25). This paper discusses in great detail the possible interactions between triazole compounds and the active site of cytochrome P-450 type enzymes at the molecular level.

In view of these results with pyrimidines and triazoles, it seems very probable that the sp^2-hybridized nitrogen atom in the cyclic azo group of the norbornanodiazetines also holds the same molecular ligand function at the active site of P-450, although no experimental results for this are available to date.

Of course, the substituents R of the individual bioregulating molecules of these nitrogen heterocycles play an additional role which is critical with regard to the macroscopically observable bioregulating effects. For example, depending on their structure, these substituents may contribute to varying, additional interactions with other, probably lipophilic, zones of the active site and in this way influence the overall interaction and binding effectivity with the target enzyme (cf. Figure 11). Moreover, these substituents R certainly influence uptake into the plants as well as translocation and metabolism, hence leading to quite different overall bioregulating effects.

In this large group of bioregulating nitrogen heterocycles, there is still plenty of room for surprises with regard to individual compounds and certainly a very wide area for empirical experimental work, both for the chemist and for the biologist.

Acknowledgments

I thank all those colleagues from chemistry and biology at BASF who assisted me in my preparations with helpful discussions and by making their results available for this review, especially Drs. E. Buschmann, K.H. König, C. Rentzea, N. Rieber, U. Schirmer, W. Türk and B. Zeeh, as well as my colleagues from biology, Drs. J. Jung and W. Rademacher.

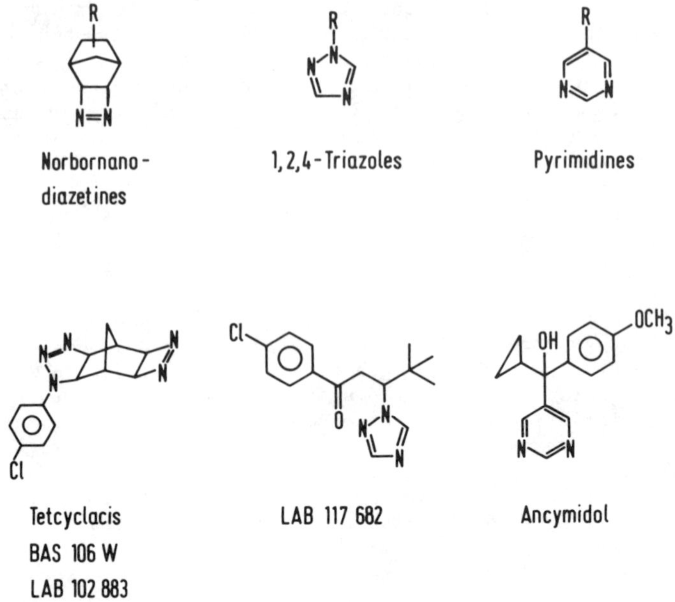

Figure 10. Three types of bioregulating nitrogen heterocycles.

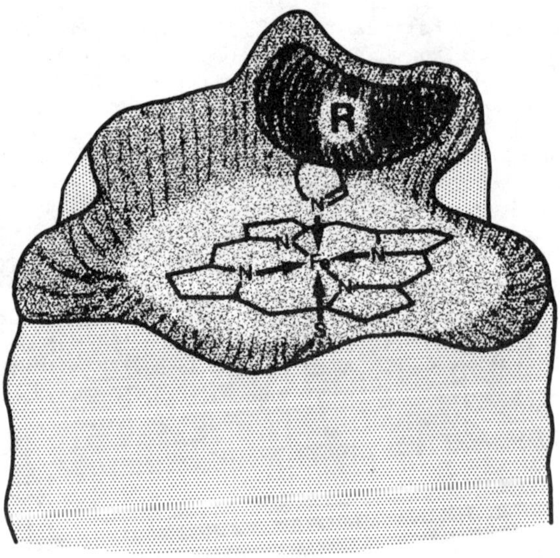

Figure 11. Model of cytochrome P-450 interacting with nitrogen heterocycle.

Literature cited

1. Tolbert, N.E. J. Biol. Chem. 1960, 235, 475.
2. Jung, J. "Plant Bioregulators in Cereal Crops: Action and Use", see chapter 5 of this book.
3. Zeeh, B; König, K.H.; Jung, J. Kem.-Kemi 1974, 1, 621.
4. König, K.H. Naturwiss. 1968, 55, 217.
5. Jung, J. Landwirtschaftl. Forsch. 1967, 20, 221.
6. Zeeh, B; Scholz, H.; Linhart, F.; Rentzea, C.; Sauter, H; Jung, J., unpublished results.
7. Rekker, R.F. "The Hydrophobic Fragmental Constant"; Nauta, W.Th.; Rekker, R.F., Eds.; PHARMACOCHEMISTRY LIBRARY Vol. 1, Elsevier: Amsterdam, Oxford, New York, 1977; p. 158.
8. Sauter, H; Jung, J., unpublished results.
9. Kier, L.B; Hall, L.H. Eur. J. Med. Chem.-Chimica Therapeutica 1977, 12, 307.
10. Rademacher, W.; Jung, J. Z. Acker- und Pflanzenbau 1981, 150, 363.
11. Sauter, H; Schirmer, U.; Türk, W.; Rademacher, W., unpublished results.
12. Hansch, C.; Leo, A. "Substituent Constants For Correlation Analysis in Chemistry and Biology", Wiley: New York, Chichester, Brisbane, Toronto, 1979.
13. Kubinyi, H., In "Progress in Drug Research"; Jucker, E., Ed.; Birkhäuser: Basel, 1979; Vol. 23, pp. 97 - 118.
14. Seydel, J.K.; Schaper, K.-J. Pharmac. Ther. 1982, 15, 131.
15. Sauter, H; Zeeh, B.; Buschmann, E.; Jung, J. Ger. Offen. 2 913 523, 1980; Chem. Abstr. 1981, 95, 24782 u.
16. Platz, R; Fuchs, W.; Rieber, N.; Samel, U.R.; Jung, J; Würzer, B. Ger. Offen, 2 615 878, 1977; Chem. Abstr. 1978, 88, 37803 g.
17. Schott, P.E; Knittel, H.; Klapproth, H. "Tetcyclacis - a New Bioregulator for Improving the Development of Young Rice Plants", see chapter 6 of this book.
18. Jung, J; Koch, H.; Rieber, N.; Würzer, B. Z. Acker- und Pflanzenbau 1980, 149, 128.
19. Rieber, N., personal communication.
20. Hildebrandt, E.; Diplom Thesis, University of Göttingen, 1982.
21. Graebe, J.E., In "Plant Growth Substances 1982"; Wareing, P.F., Ed.; Academic Press: London, 1982; pp. 71 - 80.
22. Rentzea, C.; Jung, J.; Sauter, H.; Heilen, G.; Zeeh, B. Ger. Offen. 2 739 352, 1979; Chem. Abstr. 1979, 90, 198875 p.
23. Coolbaugh, R.C.; Hirano, S.S.; West, C.A. Plant Physiol. Suppl. 1976, 59, 77.
24. Coolbaugh, R.C; Hirano, S.S.; West, C.A. Plant Physiol. 1978, 62, 571.
25. Gadher, P.; Mercer, B.C.; Baldwin, B.C.; Wiggins, T.E. Pestic. Biochem. Physiol. 1983, 19, 1.

RECEIVED February 22, 1984

3

Laboratory and Field Evaluation of Bioregulators

C. A. STUTTE

Department of Agronomy, Altheimer Laboratory, University of Arkansas, Fayetteville, AR 72701

In recent years a great deal of attention has been given to chemical control of physiological processes in plants. Scientists have joined efforts to understand plant regulation in such organizations as the Plant Growth Regulator Society of America and the British Plant Growth Regulator Group. Proceedings and other publications from such organizations illustrate and highlight the extensive and intensive work that is being done with growth regulators (1,3). Most important perhaps is the use of plant bioregulators to increase food supplies of the world. Recent literature illustrates our present capabilities for regulating crop plant growth and enhancing yields (e.g. 2).

Technological developments in bioregulation are many. We know that bioregulators are capable of altering genetic potentials by modifying biochemical and physiological processes. By altering physiological processes with bioregulators, we may be affording the plant the means to resist or overcome environmental or biological stresses and may, thereby, contribute to enhanced yields. However, it is necessary to transfer this bioregulation technology to the farmer so that his crop production and management can benefit from the technology. To make this transfer, the physiology of a specific crop and the characteristics of bioregulators need to be managed as a part of the overall system of crop management.

It is the purpose of this paper to convey to the reader the concept of using both laboratory and field evaluations to aid in developing needed information for using bioregulators in crop physiology/management. Each crop must have special considerations not only because of cultivar variations but because the interactions with the various environments must be understood.

A physiological understanding of a specific crop requires information inputs from laboratory and field experiments. Our objective has been to develop physiological laboratory methods that will provide information to plan and carry out field tests so that we may predict plant responses. This information should improve our cultural practices with positive planning and the selection of varieties and practices that will provide the best performance under specific growing conditions.

Many chemicals have been introduced into laboratory and field

0097-6156/84/0257-0023$06.00/0

evaluation programs to determine their potential utility in crop production. Most of the chemicals tested in field plots were selected on the basis of their activity in specific laboratory tests. Chemicals that do not show response in the laboratory tests are dropped from further consideration. This selective process eliminates a large number of chemicals from the more expensive and time-consuming field evaluation programs. The effective use of bioregulators in crop production may require that many current cultural practices be re-evaluated. Some new practices may have to be adapted in order to achieve maximum productivity.

Laboratory Techniques for Evaluation

Since it is very expensive and time consuming to evaluate chemicals that are available and that could be useful in crop production, it becomes necessary to develop research methods to obtain data which can rapidly identify potentially useful chemicals and provide data on how they may best be utilized.

Over the past few years much research attention has been directed to developing and utilizing laboratory methods to evaluate physiological responses to synthetic chemicals. In order to reduce the variability between plants in tests, our evaluations are made using primarily soybean tissue from single plant seed sources. This allows for less test variability by providing a high degree of genetic uniformity and also minimizing variability due to specific environments at the production site.

The following series of tests are presently being utilized for evaluation of physiological responses induced by experimental bioregulators before extensive evaluation under field conditions. Other tests may also be used. This series is used to illustrate the potential.

O_2 evolution

$^{14}CO_2$ fixation

water uptake

nitrate uptake

potassium uptake

transpiration rates

nitrogen loss from plant foliage and temperature effects on N loss

ethylene evolution from leaf tissue

phenolics and phenolic glycosides (total and composition)

field yield response.

O_2 Evolution - CO_2 Fixation

This laboratory test helps to classify a chemical as to its effect on a) increasing, b) decreasing, or c) having no effect on the $^{14}CO_2$ fixation processes involved in photosynthesis and its consequent effect on O_2 evolution processes.

Many questions can be answered with this test: a) Are the processes of CO_2 fixation and O_2 evolution dependent upon each other? b) Can the CO_2 fixation be increased while the O_2 evolution is being decreased? Using this approach, we have identified chemicals that increase $^{14}CO_2$ fixation while reducing O_2 evolution by 20 to 40 percent. This may be due to the reduced levels of O_2 internally that reduce O_2 competition with CO_2 at the site of the enzyme activity which is involved in CO_2 fixation. This would assume that CO_2 fixation is not a major limiting factor to high crop yield, but that the high level of O_2 in the tissue greatly increases oxidation of assimilates that reduce yield potentials. Therefore, one may be looking for a chemical that reduces oxidation by lower O_2 production in the tissue when in light, and this can be measured by O_2 evolution, while increasing CO_2 fixation or having no effect on the process. If a chemical inhibits both O_2 evolution and CO_2 fixation, it may be a candidate for a herbicide, or some other use as a harvest aid, but is not likely to be useful as a crop yield enhancer. When a chemical shows potential in this test, it may be further evaluated as to its effects on other processes.

Water, Nitrate, and Potassium Uptake

Chemicals that would have potential as yield enhancers would probably have little effect on reducing water and nutrient uptake by the roots. If so they would be primarily used in early vegetative growth stages or as harvest aids in such crops as sugar cane to increase sucrose production prior to harvest.

Using plants from a single plant seed source, one can evaluate the effects of chemical treatment on these processes in a controlled environment chamber. The plants (soybeans, for example) are grown in hydroponic nutrient media, and at the time of treatment (V4 or R1 stage) the nutrient media with known concentration of elements are monitored for amounts and rates of uptake of water, nitrates, and potassium. Other elements can also be monitored. In soybeans, if a chemical reduces the uptake of potassium, it is not likely to be useful for application during pod-fill; however, it may be useful during early vegetative stages to conserve soil potassium in the root zone for use during the reproductive growth stages. If a chemical reduces water uptake without greatly affecting other important physiological functions, it may be useful in increasing efficiency of water utilization. A chemical that reduces nitrate uptake but still induces greater production of dry matter may be useful in increasing nitrogen utilization efficiency with higher grain yields.

Nitrogen Loss from Plant Foliage

With the discovery that nitrogen is lost from plant foliage with transpiration and the utilization of a pyro-chemiluminescent method for analyzing the amounts of nitrogen in reduced and oxidized forms, another laboratory method has been added to collect useful data. This nitrogen dissipation that is collected in a tube placed in dry ice has been correlated directly with temperature and transpiration. The nitrogen loss is greater in senescing and stressed tissue. Chemicals may be identified that will reduce the nitrogen loss at high temperatures (35 C and greater) and increase crop yields by more efficient utilization of available nitrogen. The utilization of nitrogen in plants involves many high energy-consuming processes. If this energy is conserved rather than lost from the foliage, production certainly could be significantly increased. However, the researcher must be aware that the loss of nitrogen from the foliage with transpiration may be also a defensive process to protect the tissue from accumulating toxic concentrations that could be fatal to the plant.

Ethylene Evolution from Leaf Tissue

When most plants are subjected to stress conditions, the production of ethylene is increased. Ethylene has been shown to affect many physiological processes. Therefore, the conditions that alter ethylene production may be useful information in evaluating responses to chemicals.

Over the last four years many organic compounds have been evaluated using a rapid method that utilizes leaf disc tissue from plants having a single parent seed source. The leaf discs are submerged in the experimental test solution for approximately 20 minutes, after which they are rinsed with deionized water and rapidly blotted, then placed into a 4-cc vial, and sealed with a serum stopper. After different intervals, samples are removed for ethylene determinations. The ethylene is reported as μl ethylene/liter/cm^2/hr. This laboratory method, which is rapid and allows for many chemicals to be evaluated, has identified compounds that stimulate ethylene evolution as well as inhibit ethylene evolution. Compounds are selected that stimulated ethylene evolution in both light and dark as well as at high and low temperatures for field evaluation as stimulators of cotton boll opening. The potential for utilization of this type information about chemicals that stimulate and inhibit ethylene production are many and can be useful in many crop production practices.

Phenolics and Phenolic Glycosides

Phenolics were once considered as waste products; however, they are now recognized as very important to the growth and well-being of plants. Phenolics may be involved in the resistance phenomenon of plants to various pests. The phenolic constituents present and those produced at the time and at the site of invasion are both thought to influence pest resistance. The influence that phenolic constituents have on enzyme activities within the plant also is of great importance.

Chemicals are applied to plants grown in controlled environmental chambers from single plant seed sources. Seven days after treatment comparable leaf tissue is analyzed for phenolic composition by high pressure liquid chromatography. Chemicals identified may alter phenolic metabolism and, consequently, susceptibility or resistance to specific pests. There are many implications of the physiological-biochemical interactions as they are related to applied agriculture.

Field Test Evaluations

Correlated with laboratory tests are field test plots. Field plots are put out each year to evaluate chemical treatment responses in fields using different cultural practices. The chemicals tested in the field plots are selected because of their specific activity in the laboratory tests. Many of the supplementary observations in field studies may suggest that the effectiveness of a bioregulator will require modification of many of our cultural practices, varieties, row spacing, seeding rates, planting dates, pesticide treatments, fertilizer programs, irrigation practices, and crop rotation schedules, all of which seem to be involved in successful use of bioregulators. One must realize that, in addition, the chemicals themselves need to be further tested in order to determine the best combination in formulation, rate, time of application, spray volume, spray equipment, and surfactant or adjuvant that will result in the greatest economic return and provide environmental protection.

Summary

The advantages of subjecting a potential agricultural chemical to laboratory evaluation before field testing are many: 1) A small amount of chemical is necessary for the tests; 2) costs can be reduced by eliminating from further testing chemicals that have high probability of not working; 3) laboratory tests can identify the possible rate of the chemical necessary to bring about a specific response in field conditions; 4) these tests can help to predict the crop and the time of treatment most likely to succeed; 5) they can provide a storehouse of data on plant responses to chemicals that can be useful when additional scientific discoveries are advanced; 6) they provide chemical structure-plant response relationships that can be useful in the synthesis of chemicals to induce a desired response; and 7) these tests reduce the time and cost of product development with a higher level of safety to the public by reducing the quantity of chemicals necessary for experimentation.

The potential for utilization of bioregulating chemicals is large. However, much information is needed on physiological response before crop production can efficiently utilize a program by which physiological responses are managed with bioregulating chemicals. Growers and researchers must cease to look for miracles and search for means by which physiological processes can be enhanced or inhibited to significantly increase yields or assist in crop production. We know that the genetic potential is present in our commonly produced crop cultivars; however, this potential is not being achieved with our present cultural practices. With the use of bioregulators as physiological managers, we can greatly improve our potential for increasing efficiency of several physiological processes and increase crop yields. The potentials are great. We must not neglect the needed research that can provide expanded production around the world.

Literature Cited

1. British Plant Growth Regulator Group Proceedings, Bulletins and Monograph, 1979 through 1983.
2. Nickell, Louis G. (ed.) 1983. Plant Growth Regulating Chemicals Vol. I and II. CRC Press Inc., Boca Raton, FL.
3. Plant Growth Regulator Proceedings, Handbooks and Bulletins, 1975 through 1983. Publications of the Plant Growth Regulator Society of America.

RECEIVED April 4, 1984

Plant Bioregulators in Cereal Crops

Action and Use

JOHANNES JUNG

BASF Aktiengesellschaft, Landwirtschaftliche Versuchsstation, D-6703 Limburgerhof, Federal Republic of Germany

Of the substances with bioregulatory action (PBRs) that have so far become known, only three have been used in cereal crops to any significant extent. However, particularly in European agriculture, it has been possible to solve a big problem, namely the reduction of lodging in wheat, rye, barley and oats through treatment with growth retardants of the onium type (chlormequat chloride, CCC; mepiquat chloride, DPC) as well as the ethylene generator, ethephon.

The successful application of these PBRs leads to the question of further possibilities of optimization induced by changes in the phytohormone status of the various cereal species. There is particular interest in the favourable effect on the number of ear-bearing stems per area, the number of grains per ear, and the average grain weight. Further ways of optimizing production in cereals using PBRs can be seen in the increased stress tolerance and the improved utilization of water and plant nutrients via better root development.

The specific development of PBRs relevant to practice presupposes not only further knowledge about phytohormone control mechanisms in yield-determining factors and stress behaviour in cereal plants but also sensible projection into the particular cereal cultivation system.

In the discussion on the possibilities of using plant growth or bioregulators (PBRs) in intensive cereal cultivation, a picture is frequently presented that might better fit into the period immediately preceding the turn of the millennium. On the other hand, we already have such highly efficient production models available for cereal cultivation, and not only in research laboratories and growth chambers. These have already been put into practice in some of the areas worldwide under cereal cultivation, which is why they can be offered to other sectors of agriculture for a critical examination of their suitability under given cultivation conditions.

0097-6156/84/0257-0029$06.00/0

The yield potential of wheat is stated by Aufhammer (1) to be about 22,500 kg/ha. Thus the genetic potential of this most important cereal is by no means being fully exploited, whereas this exploitation is of course only feasible with appropriate production methods and a high outlay on operating supplies, the use of which solely makes sense when there is an adequate economic return.

The yield components that can be influenced in the cereal plant and the active compounds that are presently available for this are the subject of this contribution. The topics concerning the uses to which bioregulators have already been put and their potential applications are oriented mainly to cereal growing in Central Europe.

Yield structure and yield-determining components in cereals

The grain yield of an area results from the yield of each individual plant. The latter yield depends, in its turn, on the success with which the cereal plant passes through the various development stages. Taking these development stages into account, Figure 1 shows the yield components for the individual cereal plant and for the total plant population.

Particular importance attaches to:

- the number of shoot per plant and the density of shoots,
- the number of ears and grains per plant and area, and
- the average grain weight.

Attention must also be drawn to a phenomenon that has found particular interest in connection with the formation of yield in crop physiology: this is the reduction of the number of shoots, spikelets and florets that form at an early development stage.

It is characteristic for the development of a cereal plant that the number of shoots approaches a maximum between emergence and the beginning of stem elongation to drop again in the reduction phase as the result of competition and stress situations. The formation of ears with the number of florets per ear already determined in the tillering phase is, after quickly passing through a maximum, also subject to a reduction depending on the particular development conditions. As the third important yield component, the average grain weight is decisively influenced by the deposition of assimilates during the grain filling phase.

Possibilities of influencing yield formation with bioregulators in cereal crops

It is sufficiently well known that

- cultivation measures (in particular sowing density),
- crop nutrition measures (especially the level of nitrogen fertilization), and
- crop protection measures (mainly the elimination of competing weeds and phytopathogenic fungi)

lead to optimization of the yield components mentioned and to maximization of the overall yield. This has increasingly raised the question as to what extent substances with phytohormone or also metabolic action, which are grouped together under the term plant bioregulators, can be additionally used for exerting a positive influence on the yield of various cereal species.

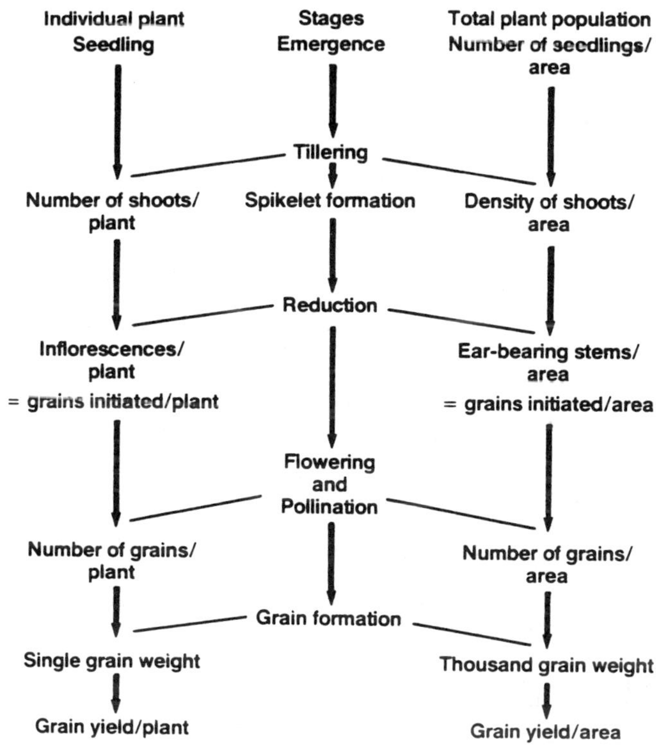

Figure 1. Yield determining components for the individual plant and for the total plant population in relation to different stages --terms and survey.

General characterization of relevant substances with a bioregulatory effect in cereal species

Classification into groups according to principle of action. Of some 60 commercially available substances with a bioregulatory effect - apart from the much larger number of active compounds described in the literature - scarcely more than three have so far actually been used in practice in cereal growing (2). To provide an overview of the most important substances that have become known so far, a classification into groups can be carried out on the basis of the principle of action.

In most cases the latter is the intended influencing of the plant's hormone status according to the following possibilities:

- supply of a phytohormone participating in the existing equilibrium or of an analogous compound;
- enhancing or inhibiting the hormone biosynthesis with exogenous compounds (precursors or synergists, and antagonists or inhibitors);
- varying availability of a phytohormone at the site of action by influencing its transport and catabolism.

On this basis, Table 1 lists some of the substances that are relevant to cereal crops, together with the phytohormones known so far considering their relation in terms of action of the latter.

At the present time, the compounds interfering with gibberellin biosynthesis (GA antagonists) and ethylene generators are worth of special interest because they have already found extensive application in some areas of world cereal cultivation. In the case of the other groups, interest is concentrated on the question of their potential use.

Mechanisms and spectra of action. An indication of the mechanism of action of the various bioregulators has already been given with their classification in Table 1. However, the gibberellin antagonists will be dealt with in a little more detail; on the one hand, on account of their importance in practice, but, on the other hand, because we have been working on them for some time.

For GA antagonists of the onium type, which include CCC (chlormequat chloride) and DPC (mepiquat chloride) as growth retardants, it is assumed that they inhibit the cyclization of geranylgeranyl pyrophosphate to copalyl pyrophosphate in the course of GA biosynthesis (survey see 3). Ancymidol, tetcyclacis (NDA) and triazoles inhibit the sequential oxidation of ent-kaurene to ent-kaurenoic acid (4, 5, 6). These relationships are shown in Figure 2.

It must be stated in connection with the spectrum of action of the GA antagonists that onium compounds have a declining intensity of action from wheat, via rye, barley, oats to rice and maize. Tetcyclacis, however, after being taken up by the roots, has an approximately comparable effect in all cereal species. An example of the effect on rice of the various substances mentioned can be seen in Figure 3. Particular attention should be drawn to the pronounced difference in action between the two groups of cyclization and oxidation inhibitors of GA biosynthesis.

Table I. Overview of relevant substances with a bioregulatory effect in cereals classified according to their phytohormonal interaction.

Reference phyto-hormone	Compounds with homologuous or synergistic activity	Compounds with an antago-nistic activity or inhibitors of biosynthesis
Auxins	Synthetic Auxins (NAA, 2,4-D . .)	Triiodobenzoic acid (TIBA)
Gibberellins	Steroids	Onium compounds (CCC, CMH, DMC, DPC)
	Phthalimides	Pyrimidine derivatives (Ancymidol)
	Steviol	Norbornenodiazetine deriv-atives (NDA = Tetcyclacis)
		Triazoles
Cytokinins	Benzyl- and Furfuryl-aminopurine Phenylurea derivatives	
Abscisins	Terpenoic analogues of ABA Farnesol Jasmonic acid	
Ethylene	Chloroethyl phosphonic acid (Ethephon) Aminocyclopropane carboxylic acid (ACC) (Auxins)	Aminoethoxivinyl glycine (AVG)

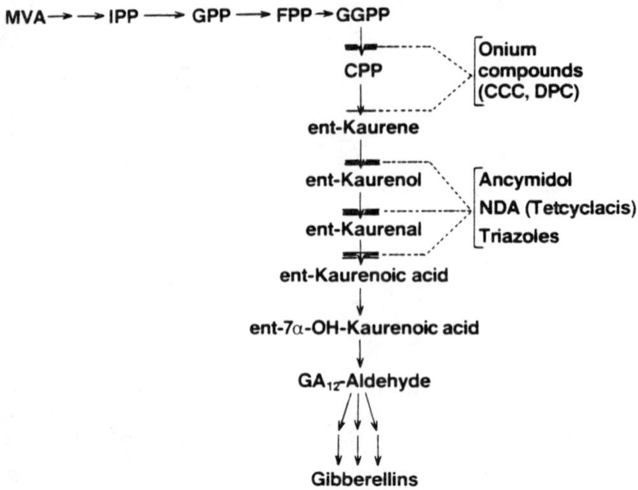

Figure 2. Interference of some growth retardants with gibberellin biosynthesis. (Abbreviations: MVA, mevalonic acid; IPP, isopentenyl pyrophosphate; GPP, trans-geranyl pyrophosphate; FPP, transfarnesyl pyrophosphate; GGPP, trans-geranylgeranyl pyrophosphate; CPP, copalyl pyrophosphate).

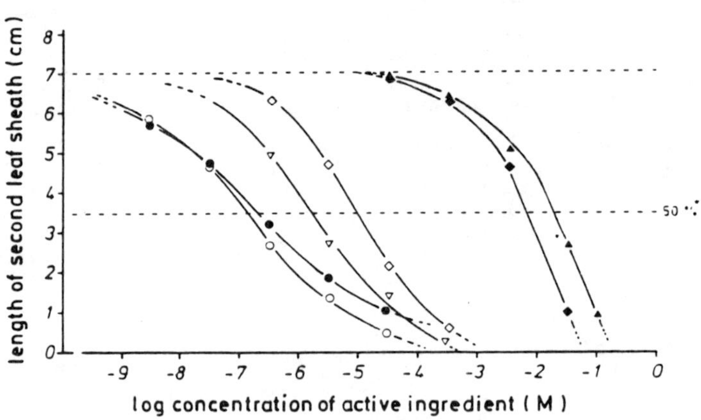

Figure 3. Action of various concentrations of growth retardants (GA antagonists) on the growth of rice seedlings. ▲–▲ = CCC; ◆–◆ = DPC; ◇–◇ =triazole 117 682; ▽–▽ = triazole 130 827; ●–●–● = ancymidol; ○–○–○ = tetcyclacis (NDA). KI_{50} = concentration of active ingredient by means of which a shortening of 50% is achieved (see also 11).

Aspects of the use of bioregulators in the various development stages of cereal species

Germination and seedling development. In the series of various development stages of the cereal plant up to the realization of grain yield, the question of influencing them appropriately by the use of bioregulators already arises at the stage of germination (7). It is known that gibberellins stimulate the germination of all cereal species vigorously, which is why, for example, GA_3 is used in the production of malt. However, there have not up to now been any significant possibilities of application in cereal crops for this type of regulation. On the other hand, substances with an opposite action to gibberellin have met with considerably greater interest both experimentally and in practice. In this connection attention must be drawn, in particular, to the seed treatment of winter wheat with CCC, as is practised in some regions to increase the winter resistance of wheat seedlings. The improved overwintering of wheat treated with CCC was already shown by us in model trials in the mid-sixties (8). The treatment of both the seed and the soil with CCC produced a clear growth advantage for the treated plants up to the end of tillering in spring. Concerning the use of CCC as a seed treatment, Zadoncev et al. (9) emphazise the lowering of the tillering node after seed treatment with CCC as the reason for improved overwintering.

There is apparent justification for regarding the treatment of seed with bioregulators having GA antagonistic action as a possibility for exerting a significant influence on the development and stress physiological behaviour of cereal plants. In particular, newer compounds such as tetcyclacis (10, 11) can cause a change in the shoot: root ratio. According to the model trials that we have carried out so far, the change in the shoot:root ratio can be achieved using these substances not only by a reduction of both organ parts varying in degree but also by an absolute promotion of root growth with a simultaneous reduction of shoot length. The example given in Figure 4 illustrates this effect. It may be assumed that these compounds can be used to optimize water and nutrient utilization and thus to optimize yield in cereal crops.

Tillering. Tillering assumes special importance within the vegetative development of a cereal plant. On the one hand, the number of tillers developed determines the density of the total crop to a decisive extent. On the other hand, even during the tillering stage, the differentiation processes for the formation of the generative organs begin, this fixing, for example, the number of spikelets in the individual ear. What is decisive for the number of ear-bearing stems per area and for the number of grains per ear is the quantitative result of the formation of these yield components and of their reduction during the subsequent phase of stem elongation.

It is known that daylength, temperature and supply of nutrients have a significant influence on the course of these processes. Less information is available, however, about the phytohormone coordination involved and the possibility of utilizing it in practice by using appropriate active substances.

Apical dominance, which is determined primarily by the auxin level, but not solely by it, no doubt has an important influence on tillering (12). Therefore, the latter can be promoted by substances having an auxin-antagonistic action, such as triiodobenzoic acid (TIBA) (13, 14). Plant bioregulators with a GA-antagonistic effect also have a stimulating influence on tillering or the number of ear-bearing stems per area, whereas gibberellin treatment reduces this (15).

Gibberellin treatment tends to hasten development, to shorten the intervals between particular morphogenetic events and to reduce the number of organs laid down in successive phases of development, whereas retardants generally have the opposite effect.

Standing ability and lodging. After stem elongation and heading the cereal plant reaches a very critical phase for the formation of yield, namely grain filling. The effective and longest possible functioning of the assimilating organs is essential. The assimilates required for grain filling are produced during a relatively short period. According to Evans (16), only 5 to 10% of the total assimilates deposited in the grain originate from the pre-flowering period. Any disturbance in the photosynthetic activity of the assimilating plant organs in this phase of development - insufficient insolation, lack of water and nutrients, lodging or parasitic attack - is bound to have an adverse effect on grain formation and yield (17).

Lodging is one of the principal problems in intensive cereal cultivation that has an unfavourable effect, particularly when it occurs in the early reproductive stage of development. On the basis of fundamental investigations carried out by Tolbert (18, 19) on the action of various quaternary ammonium compounds on the growth of wheat, it was possible to develop chlormequat (CCC) into a very effective anti-lodging agent and to introduce it to large areas of wheat growing all over the world (20, 21). In West Germany, by far the greatest part of area under wheat cultivation has been treated with CCC for many years now. Here, as in a number of other European countries, this plant bioregulator is an integral element of the wheat production system.

As a gibberellin antagonist, CCC increases the standing ability of wheat, on the one hand by reducing the stem length - especially in the lower internodes - and, on the other hand, by increasing the stem's diameter and by strengthening its wall (Figure 5).

All the wheat varieties which have so far been tested reacted to this PBR, although with differences in degree. Generally the short-straw varieties react more vigorously than the long-straw ones. In order to achieve an optimum effect, the rate of application must be adjusted to the variety and, in addition, the growth stage and the local cultivation conditions must be taken into consideration. Only this careful harmonization can exploit the possibilities which this growth regulator offers in wheat growing for increasing yield, especially on the basis of a higher supply of nitrogen.

In other cereal species - particularly in the very lodging-susceptible winter barley - the stem-stabilizing effect of CCC is considerably weaker. A search has therefore been carried out for further substances. It has emerged that, in addition to CCC, a large number of different compounds with a charged central atom (onium compounds) also have a bioregulatory effect. Some of the compounds of this class of substances are given in Figure 6. However, in spite of an increased

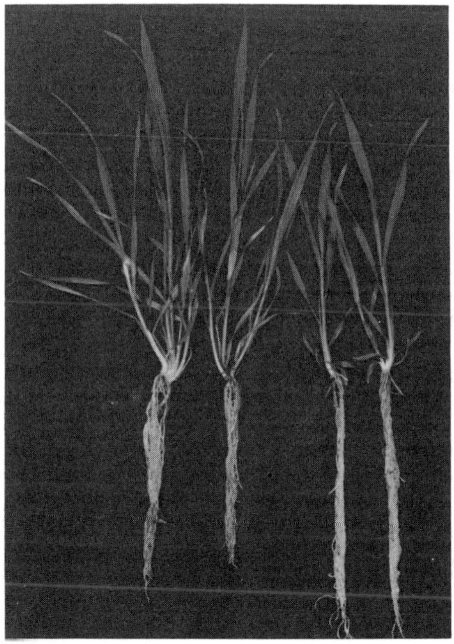

Figure 4. Example for the change of the shoot:root ratio after treatment of wheat with tetcyclacis (NDA). Left: untreated; right: 15.0 mg a. i./100 kg seeds.

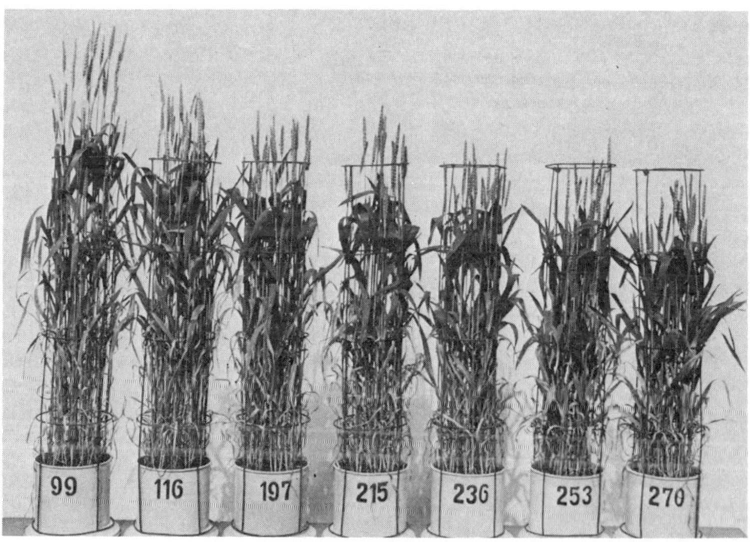

Figure 5. Effects of increasing CCC rates (0.1-3 mg/pot) on the growth of wheat.

effect in some cases compared with CCC, the intensity of stem shortening and stabilizing in winter barley is not quite achieved. Therefore, combinations of onium compounds with the ethylene generator ethephon were investigated (22, 23, 24). The two compounds complement each other. The ethylene generator ensures particularly that stem shortening continues into the generative phase, since the considerable reduction in growth height achieved by the onium compounds in the vegetative stage is often compensated later by the barley plant to a surprising degree. Ethephon can be used as a stalk stabilizer in barley not only in combination but also on its own (25).

Ear and grain development. The ear structure of cereals, especially that of wheat, exhibits differences due both to variety and development. In particular, the fact that the number of fertile florets in the differently inserted spikelets in the ear can vary as can the weight of the individual grains deserves special attention with reference to the formation of yield. According to Aufhammer and Bangerth (26), the proportion of florets that form grains varies between 40 and 80% within the ear, the tendency decreasing upward and downward from the central region (Figure 7). At the base of the ear one to four spikelets are often sterile. The weight of the individual grains within individual spikelets may fluctuate between 30 and 60 mg. Within a spikelet, the basal florets generally develop bigger or heavier grains than the distal ones.

There is much to suggest that these differences in ear and grain development are controlled by phytohormones, apparently along the lines of a dominance principle. Within an inflorescence, the oldest "dominant" floret influences the development of each subsequent one (primigenial dominance).

It is therefore not surprising that GA antagonists such as CCC and ancymidol reduce the competitive action of various sinks and improve the synchrony of ear and grain formation (27).

Besides this correlative influence, attempts are being made to promote grain formation by increasing the sink capacity. Michael and Seiler-Kelbitsch (28) have already established connections between weight per grain and the cytokinin content of cereals. However, the exogenous supply of cytokinins in the grain formation phase, which has been investigated by Herzog and Geisler (29), has so far shown only limited effectiveness, which may be linked to the transport behaviour of the phytohormone applied.

It is, however, possible that compounds of the cytokinin type and other substances having a senescence-retarding effect indirectly have a positive influence on grain filling, in that they extend the functioning of the parts of the plant that are mainly assimilating in this phase (30). On the other hand, any specific promotion of the assimilation rate by optimizing the fixation of CO_2, as has been formulated by various authors as the aim for C_3 plants (31, 32, 33), is probably still far from realization in actual practice.

Bioregulators in the cereal cultivation concept - prospective assessment

Apart from the physiological and technical aspects, an important economic fact must be taken into account in the use of bioregulators in cereal crops, which is already being practised or should be subjected

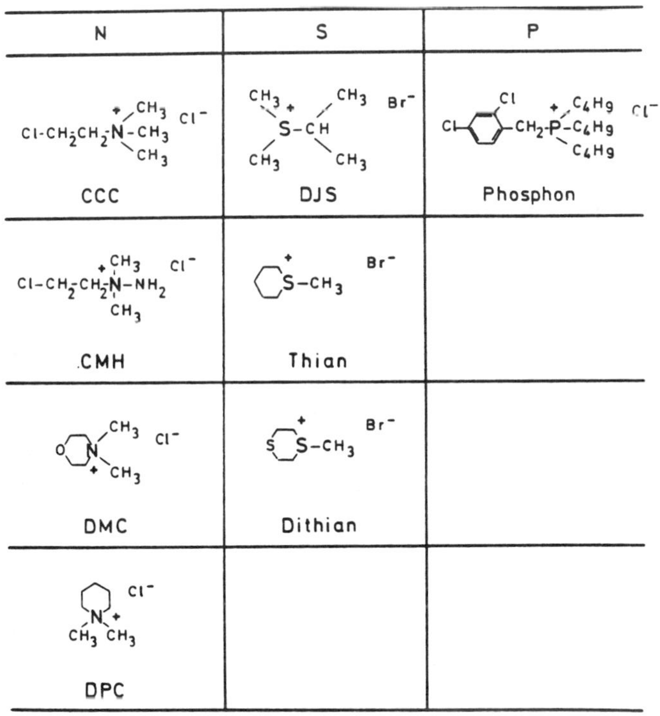

Figure 6. Bioregulators with a charged central atom (onium compounds).

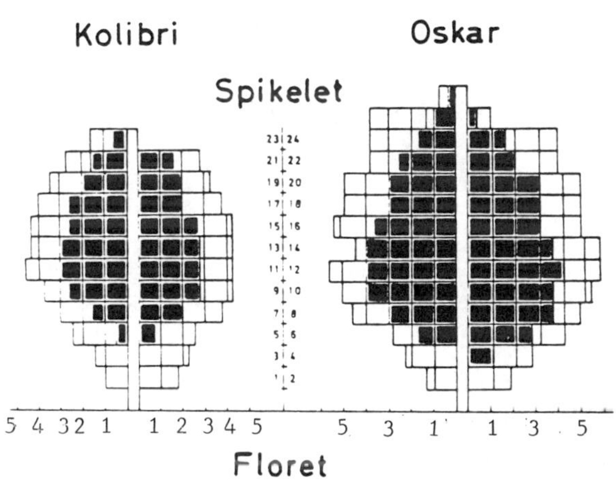

Figure 7. Number of fertile and sterile florets per ear of spring wheat varieties. Reproduced with permission from Ref. 26. Copyright 1982 Butterworth Scientific & Co. Ltd.

to a prospective assessment. This is the fact that the expenditure on production in this type of crop is normally restricted in comparison with so-called cash crops. Nevertheless, in intensive European cereal cultivation, the use of anti-lodging agents has proved profitable in combination with herbicides and fungicides and an appropriate level of nitrogen fertilization. In many cases, this profitability is probably linked to the combined effect of all measures and substances involved. Thus bioregulators may appropriately be regarded as a new generation of agrochemicals after fertilizers and plant protection agents, the efficiency and profitability of which is chiefly determined by the successful use of all products involved.

How these different measures and treatments, adapted to the crop's course of development, are projected into the production concept for cereals is shown in Figure 8 by means of a schedule which is representative of wheat growing in Central Europe. Besides this multifactorial inclusion of a plant bioregulator in the concept of agrochemical measures, there may, however, be a use that is substantially independent of this. This includes the improved overwintering of wheat in a continental climate after a treatment with CCC and the growth control of rice seedlings with GA antagonist bioregulators such as tetcyclacis. These examples of applications for bioregulators and the part of the range of this group of substances that has not yet been used, together with the list of desirable optimizations in cereal growing finally lead to the question of the future focal points of development in this field. Besides the further exploitation of the potential of bioregulators already in use, the following list of priorities may be drawn up:

1. Optimized formation of yield-determining organs of the cereal plant
   > by means of improved development in the vegetative phase resulting in an increased number of ear-bearing stalks per area;
   > by means of improved synchrony of the differently inserted spikelets and florets in the ear resulting in an increased number of grains per ear;
   > by means of promoting grain formation and the weight per grain by an appropriately intensive and sufficiently long phase of assimilate deposition resulting in a higher harvest index and improved grain weight.
2. Improved adaptation of the cereal plant to the environmental conditions; as a result, better utilization of light, water and nutrients as well as increased stress tolerance.
3. Improved resistance to pathogens.
4. Direct or indirect influence on ingredients or quality features of the cereal grain.

To what extent it will be possible to supplement the range of compounds, which is necessary to realize this list of bioregulator requirements, will depend not only on further knowledge about the phytohormonal control of yield formation in cereal crops but also and particularly on the innovation and techniques of the synthesizing chemist and experimental biologist.

Acknowledgments

My thanks are due to Mr. B.E. Byrt for the translation of the manuscript and to Mrs. B. von Hohnhorst and Mr. G. Spoor for preparatory assistance.

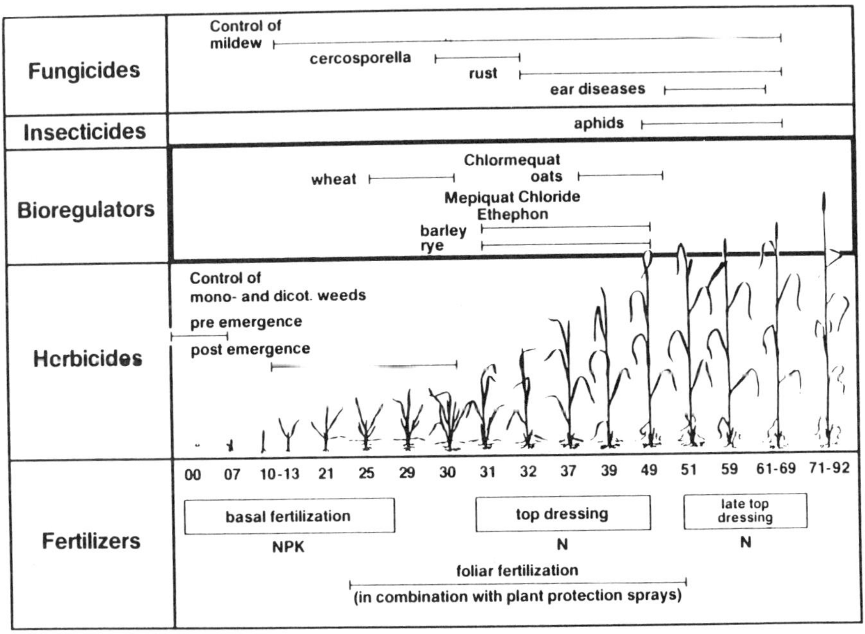

Figure 8. Application schedule for agrochemicals in cereal production.

Literature Cited

1. Aufhammer, W. DLG-Mitteilungen 14 1976, 91, 780-783.
2. Nickell, L.G. "Plant Growth Regulators - Agricultural Uses"; Springer: Berlin-Heidelberg-New York, 1982.
3. Dicks, J.W. In "Recent Developments in the Use of Plant Growth Retardants"; Clifford, D.R.; Lenton, J.R., Eds.; British Plant Growth Regulator Working Group, MONOGRAPH N° 4, Wessex Press, Wantage, 1980; pp. 1-14.
4. Coolbaugh, R.C.; Hamilton, R. Plant Physiol. 1976, 57, 245.
5. Graebe, J.E. In "Plant Growth Substances"; Wareing, P.F., Ed; Academic: London-New York, 1982; pp. 71-80.
6. Hildebrandt, E. Diplom-Dissertation, University of Goettingen, Fed. Rep. of Germany, 1982.
7. Bewley, J.D.; Black, M. "Physiology and Biochemistry of Seeds; I. Development, Germination and Growth"; Springer: Berlin-Heidelberg-New York, 1978.
8. Jung, J. Z. Acker- und Pflanzenbau 1965, 122, 9-14.
9. Zadoncev, A.I.; Pikus, G.R.; Grincenko, A.L. "Chlorcholinchlorid in der Pflanzenproduktion"; VEB Deutscher Landwirtschaftsverlag: Berlin, 1977.
10. Jung, J.; Koch, H.; Rieber, N.; Würzer, B. Z. Acker- und Pflanzenbau 1980, 149, 128-136.
11. Rademacher, W.; Jung, J. J. Agronomy and Crop Science 1981, 150, 363-371.
12. Michael, G.; Beringer, H. In "Physiological Aspects of Crop Productivity"; International Potash Institute: Bern, 1980; pp. 85-116.
13. Jewiss, O.R. Journal of British Grassland Society 1972, 27, 65-82.
14. Ackerson, R.C.; Chilcote, D.O. Crop Science 1978, 18, 705-708.
15. Koranteng, G.O.; Matthews, S. In "Chemical Manipulation of Crop Growth and Development"; McLaren, J.S., Ed.; Butterworth Scientific: London, 1982; pp. 343-357.
16. Evans, L.T. In "Crop Physiology"; Evans, L.T., Ed.; Cambridge University Press, 1980; pp. 327-355.
17. Jung, J. In "Plant Regulation and World Agriculture"; Scott, T.K., Ed.; Plenum Press: New York and London, 1979; pp. 279-307.
18. Tolbert, E. J. Biol. Chem. 1960a, 235, 475-479.
19. Tolbert, E. Plant Physiol. 1960b, 35, 380-385.
20. Linser, H.; Kühn, H. Z. Pflanzenern. Düng., Bodenk. 1962, 96, 231-247.
21. Jung, J.; Sturm, H. Landw. Forsch. 1964, 17, 1-9.
22. Kühn, H.; Schuster, W.; Linser, H. Z. Acker- und Pflanzenbau 1977, 145, 22-30.
23. Behrendt, S.; Schott, P.E.; Jung, J.; Bleiholder, H.; Lang, H. Landw. Forsch. 1978, SH 35, 277-290.
24. Herbert, C.D. In "Chemical Manipulation of Crop Growth and Development"; McLaren, J.S., Ed.; Butterworth Scientific: London, 1982; pp. 315-327.
25. Hill, D.M.; Joice, R.; Squires, N.R.W. In "Chemical Manipulation of Crop Growth and Development"; McLaren, J.S., Ed.; Butterworth Scientific: London, 1982; pp. 391-397.
26. Aufhammer, W.; Bangerth, F. In "Chemical Manipulation of Crop

Growth and Development"; McLaren, J.S., Ed.; Butterworth Scientific: London, 1982; pp. 359-373.
27. Höfner, W.; Kühn, H. In "Chemical Manipulation of Crop Growth and Development"; McLaren, J.S., Ed.; Butterworth Scientific: London, 1982; pp. 375-390.
28. Michael, G.; Seiler-Kelbitsch, H. Crop Sci. 1972, 12, 162-165.
29. Herzog, H.; Geisler, G. Z. Acker- und Pflanzenbau 1977, 144, 230-242.
30. Grossmann, K.; Jung, J. Z. Acker- und Pflanzenbau 1982, 151, 149-165.
31. Zelitsch, I. In "Encyclopedia of Plant Physiology", Vol. 6; Gibbs, M.; Latzko, E., Eds.; Springer: Heidelberg-Berlin-New York, 1979; pp. 353-367.
32. Keys, A.J.; Bird, I.F.; Cornelius, M.J. In "Chemical Manipulation of Crop Growth and Development"; McLaren, J.S., Ed.; Butterworth Scientific: London, 1982; pp. 39-53.
33. Treharne, K.J. In "Chemical Manipulation of Crop Growth and Development"; McLaren, J.S., Ed.; Butterworth Scientific: London, 1982; pp. 55-66.

RECEIVED April 4, 1984

5

Tetcyclacis: A New Bioregulator for Improving the Development of Young Rice Plants

PETER E. SCHOTT and HARRY KNITTEL

BASF Aktiengesellschaft, Landwirtschaftliche Versuchsstation, D-6703 Limburgerhof, Federal Republic of Germany

HARALD KLAPPROTH

BASF Agricultural Research Farm Taipei, Taipei, Taiwan

Studies have demonstrated the new bioregulator Tetcyclacis (BAS 106 04 W) to be suitable for programming rice seedling propagation. The substance inhibits oxidative reactions from ent-kaurene to ent-kaurenoic acid and consequently the production of biological active gibberellins. Treated rice seedlings are more compact and smaller than untreated ones.

By shortening the length of the 1st and 2nd leaf sheaths as well as the plant height, the seedling quality could be kept up for a longer period. Such seedlings therefore are more resistant to transplanting stress. Furthermore, treated seedlings show improved rooting activity. The leaves of treated plants show a more upright standing at and after transplantation. The production of dry matter is increased.

Most suitable is a seedsoaking application. Application rates from 5-15 ppm (approx. 0.25 to 0.76 g/ha a.i.) are sufficient.

Today approximately 75% of the world's rice area is grown as paddy rice (1). Asia is the principal rice growing area. The wet rice is raised mainly from transplanted seedlings. Strong and healthy seedlings transplanted at a proper time are one of the conditions for high yield. Generally the transplantation is done 9-14 days after seeding into the seedbed, e.g. Dapog system on the Philippines, and 20-30 days after seeding in boxes, e.g. Japan and China.

0097-6156/84/0257-0045$06.00/0

Often it happens that, for example due to unfavourable weather conditions, transplantation of seedlings is delayed. Under these circumstances the seedlings are growing out of control. Such weak plants are too tall and sensitive towards transplanting stress. Also the use of transplanting machines is limited.

As the mechanisation of transplanting rice is continuously increasing the economy of rice growing will get questionable. For the time being, rice is mechanically transplanted in

Japan by approx.	85% of the total rice area (i.e. approx. 2.3 Mio. ha)
Taiwan	75% of the total rice area (i.e. approx. 0.4 Mio. ha)
China	20% of the total rice area (i.e. approx. 7.2 Mio. ha)
Korea, South-	10% of the total rice area (i.e. approx. 0.12 Mio. ha)

20-21 days after seeding a proper young seedling should be not taller than approx. 13-15 cm; otherwise mechanical transplantation is impaired. Therefore the purpose of our bioregulator research work was to find a substance effective for the control of overgrowth of rice seedlings before transplantation.

Since Tetcyclacis was discovered (3, 5) extensive studies have been done on rice seedlings. Positive results on a temporary growth inhibition can be explained as a specific inhibition of oxidative reactions from ent-kaurene to ent-kaurenoic acid (2) (Figure 1).

Materials and methods

Tetcyclacis (common name accepted by ISO; BASF Aktiengesellschaft, D-6700 Ludwigshafen, Federal Republic of Germany), active ingredient of KENBYO, registered trade mark of BASF Aktiengesellschaft, formulated as a 1% soluble powder was tested in microplots as well as under field conditions, alone and in comparison with other established bioregulators (Figure 2).

Application rates were 0; 5; 6; 7.5; 10; 15; 50 and 100 ppm of active ingredient. Tetcyclacis was applied either as a seedsoaking treatment at 25-30°C for 24 hours, as dipping treatment for 30 seconds after dormancy breaking or as a drench treatment during growth stages according to Zadoks scale (10) 10 and 11.

The rice varieties included in the tests are shown in Figure 3.

In the trials a four times replicated and randomized block design was used. Depending on the trial some or all of the following parameters were evaluated: plant height; length of 1st and/or 2nd leaf sheath respectively; number of true leaves per plants; dry weights of roots and shoots and grain yield.

All trial results obtained were calculated by means of analysis of variance. Comparision of different treatments has been done by using the Duncan or Tuckey test respectively.

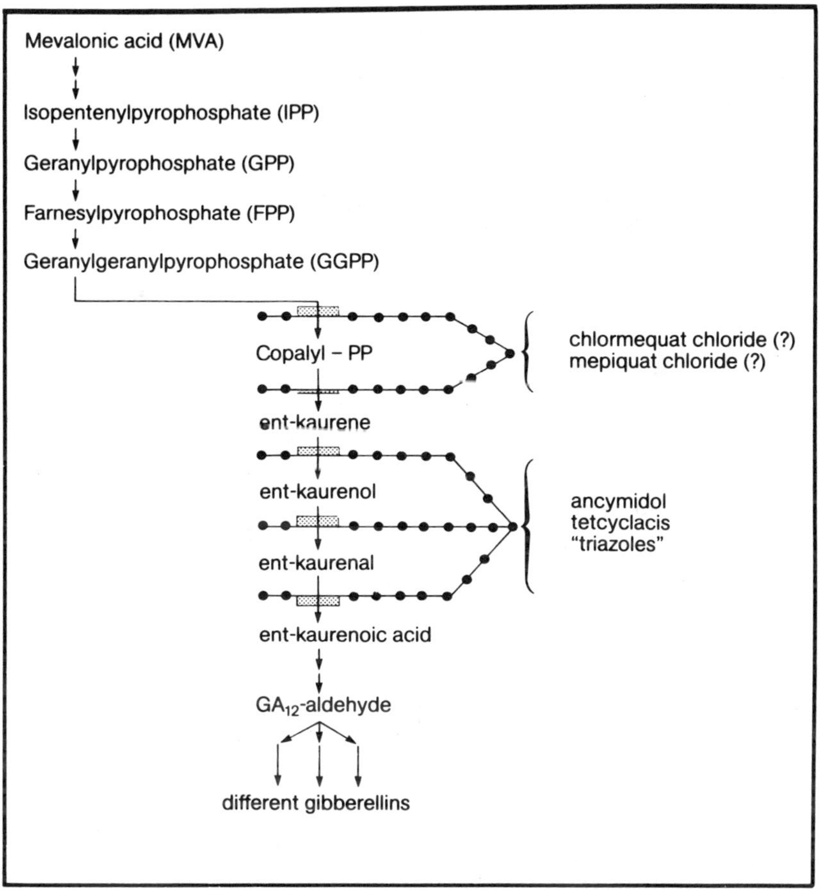

Figure 1 Pattern of Gibberellin biosynthesis and points of inhibition by Tetcyclacis

formula	common name	chemical name
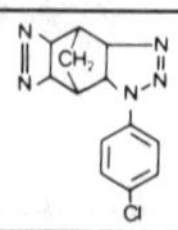	*Tetcyclacis (accepted by ISO)	5-(4-chlorophenyl)-3,4,5,9,10-pentaaza-tetra-cyclo [5,4,1,$0^{2,6}$, $0^{8,11}$] dodeca-3,9-diene
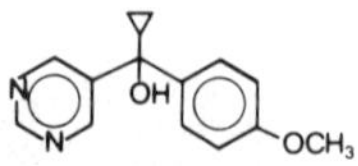	Ancymidol	α-cyclopropyl-4-methoxy-α-(pyrimidin-5-yl) benzyl alcohol
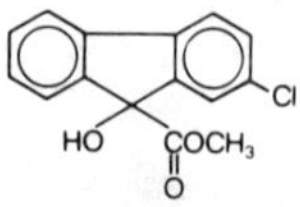	Methyl chlor-flurenol	methyl-2-chloro-9-hydroxyfluorene-9-carboxylate
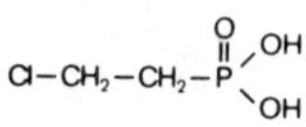	Ethephon	(2-chloroethyl)-phosphonic acid
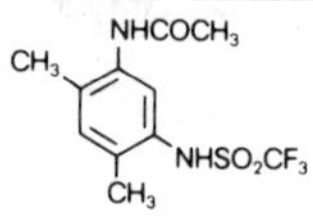	Mefluidide	N-[2,4-dimethyl-5-[[(tri-fluoromethyl) sulfonyl] amino] phenyl] acetamide
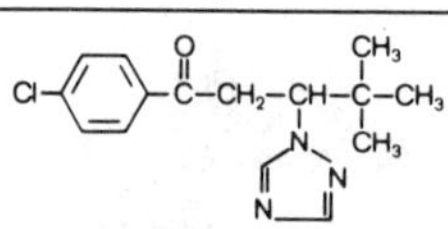		(4-chlorphenyl)-[3,3-dimethyl-2-(1,2,4-triazol-1-yl)butyl]-ketone
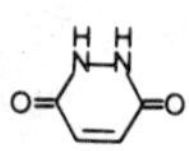	Maleic hydrazide	1,2-dihydro-3,6-pyridazinedione
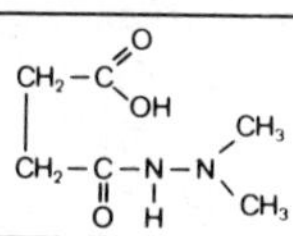	Daminozide	succinic acid,2,2-dimethylhydrazide
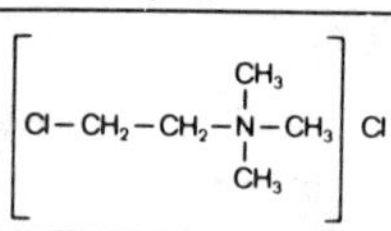	Chlormequat chloride	2-chloroethyl-trimethylammonium chloride
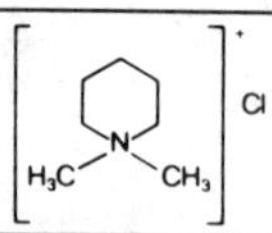	Mepiquat chloride	N,N-dimethylpiperidinium chloride

*common name (accepted by ISO), BASF Aktiengesellschaft, D-6700 Ludwigshafen, Federal Republic of Germany

Figure 2 Bioregulators compared to Tetcyclacis

variety	group
Allorio	javanica
Arborio J 1	javanica
Arborio J 10	javanica
Arditane	javanica
Lomelto	javanica
Skirpe 136 anthocyan	javanica
Ballila	indica
Bluebelle	indica
Boshito	indica
Panbira	indica
Inabawase	japonica
Koshihikari	japonica
Nihonbare	japonica
Tainung 67	japonica
Milyang 23	indica x japonica
Yushin	indica x japonica

Figure 3 Rice varieties on which Tetcyclacis has been tested

Results

Rate and time of application. A comparison of various rates and application time leads to the conclusion that with increasing the rate of Tetcyclacis the growth inhibition of plant height and 1st leaf sheath will also increase significantly (Figure 4). Sufficient are application rates between 5 and 15 ppm, i.e. by using 1.8 l water/kg seed; 113 g seeds/seedling box and 250 boxes/ha a final application rate of approx. 0.25-0.76 g/ha.

Strongest effects are obtained out of treatments as seedsoaking and dipping shortly after dormancy breaking. The later the treatment the smaller the efficacy. Dry weight of roots and shoots, however, was improved by all treatments. It is interesting to note a change of the allometry (9) only after seedsoaking treatment with Tetcyclacis to the advantage of an increased root production. A better root system on treated plants is directly correlated with a better development of the crop.

Reduced height growth due to the treatment with Tetcyclacis may be used for higher seeding densities in the riceboxes. An increase of approximately 10-20% seems to be practicable. There is no negative effect on plant growth or on the number of leaves per plant. An increase of the seeding density will allow one to save capacity of seedboxes and soil in the seedling center. It is typical for such treated plants to show a decreased plant height but also more leaves per plant (Figure 5).

Treated rice plants are more compact and shorter in length than untreated ones. This permits a delay in the date for transplanting into the field by several days (Figure 6).

A similar growth curve has been found with the indica rice type, cv. Ballila from Spain. Again due to the seedsoaking treatment, the date of transplanting could have been delayed by approx. 6 days compared with untreated plants from watersoaked seeds (Figure 7).

Since the inhibition effect of a seedsoaking treatment on the plant height will decline until approximately 20-30 days after seeding a stimulation of the total plant growth soon after transplantation is observed (Figure 8). The increase in growth is given in percent compared to the measurement of the previous growth stage,

$$(\text{growth rate} = \frac{\text{plant height at d + 1}}{\text{plant height at d}} \times 100;\ \text{d = date of measurement}).$$

From the beginning of the 2nd true leaf stage (Zadoks 12) the treated plants are showing an increased growth rate of approx. two times as much as the control seedlings.

Analysing a rice plant (Figure 9) approx. one month after seedsoaking the plant height reduction disappears more and more compared with an untreated plant from watersoaked seeds. The first and second leaf sheaths, however, still remain significantly shorter than on the control plants. This will ensure an upright standing also after transplanting in to the field. That means, the leaves will not touch the water surface which often could still be too cold and retard the growth of fresh transplanted rice seedlings. According to Tanaka (7) a better light reception of such leaves on both sides and the improve leaf formation will lead thus to maximal CO_2-absorption.

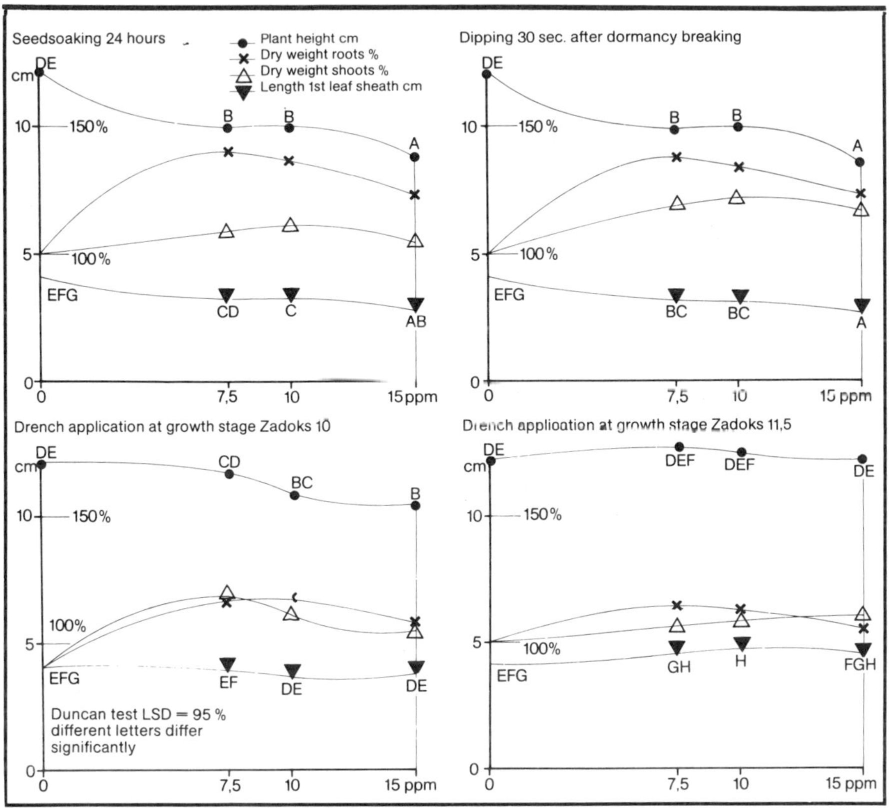

Figure 4 Interaction of Tetcyclacis vs. application time on Oryza sativa, cv. Tainung 67. - Evaluation of measurement 22 and weight 38 days after treatment -

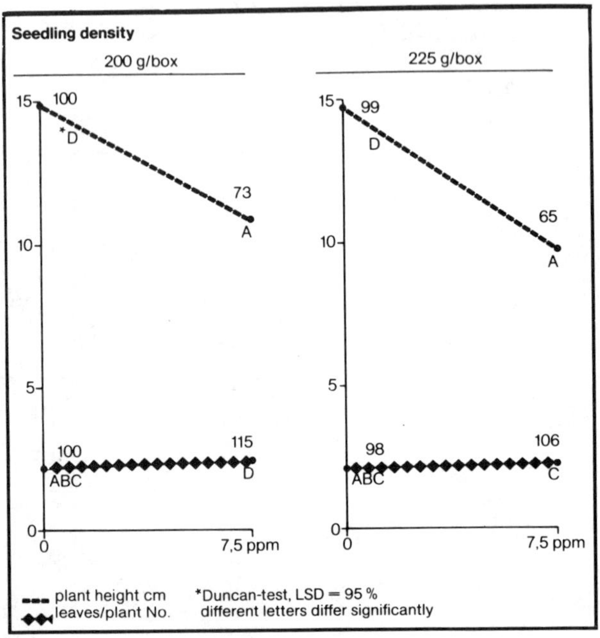

Figure 5 Interaction of seedsoaking treatment with Tetcyclacis vs. seeding density of Oryza sativa, cv. Tainung 67. - Evaluation 34 days after treatment -

growth stage	Tetcyclacis ppm		ratio
	0	10	
Zadoks 11	22,9%	21,9%	1:1,0
Zadoks 12	4,6%	9,8%	1:2,1
Zadoks 13	6,2%	8,9%	1:1,4
Zadoks 14	2,5%	9,1%	1:3,6
21st–31st day after seeding	4,4%	9,3%	1:2,1

Figure 8 Influence of Tetcyclacis on the relative growth rate of seedsoaked Oryza sativa, cv. Ballila (indica) seedlings compared with control seedlings from watersoaked seeds

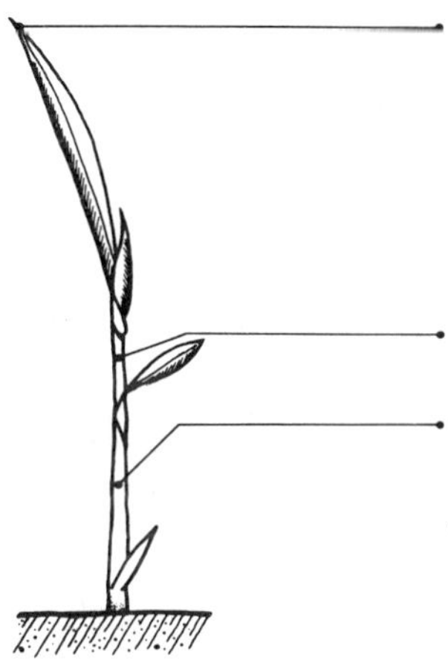

		Tetcyclacis ppm 0	Tetcyclacis ppm 10
(21) plant height	cm %	12,0 100 B**	8,0 61 A
(27) plant height	cm %	17,2 100 B	13,3 77 A
(31) plant height	cm %	19,7 100 A	18,4 93 A
(33) No. of leaves (Zadoks)		14,0	14,3
(33) 2nd leaf sheath length	cm %	8,5 100 B	5,9 70 A
(33) 1st leaf sheath length	cm %	5,4 100 B	3,4 64 A

**Tuckey-test, LSD 95%

*(21) plant height	cm	1,34
*(27) plant height	cm	1,07
*(31) plant height	cm	1,66
*(33) 1st leaf sheath length	cm	0,34
*(33) 2nd leaf sheath length	cm	0,55

different letters differ significantly

*days after seeding in pots.

Figure 9 Influence of a Tetcyclacis seedsoaking treatment on plant growth of a young Oryza sativa, cv. "Ballila" (indica) seedling

Varietal response on Tetcyclacis. In one test done at Ebina, Japan with 14 different varieties, the efficacy of Tetcyclacis on possible specific varietal reaction was tested. None of the varieties showed phytotoxic symptoms after treatment.

Two weeks after seedsoaking in a 10 ppm Tetcyclacis solution the rice varieties of the indica type showed the strongest height reduction with 13% followed by the indica x japonica varieties with 12%. At the same time the japonica and javanica varieties showed an average plant height reduction of 10% respectively 9%.

On the 19th day after seedsoaking the strongest height reduction of 13% was obtained by the japonica varieties, followed by the javanica ones with 11%. The height reduction effect nearly disappeared on the indica x japonica and the indica varieties with 3% and 2% respectively.

At the same time the 2nd leaf sheath was reduced the most by 24% on the japonica and javanica rice varieties. However the indica and indica x japonica varieties still showed 17% and 16% reduction, respectively.

By increasing the application rate from 10 to 50 ppm the above mentioned effects could be extended for plant height and length of the 2nd leaf sheath by 12-14% on the japonica and indica x japonica varieties and by 6-12% on the javanica and indica varieties. The strongest effect on overall growth inhibition at transplantation time was found within the varieties of the japonica and javanica group, whereas the varieties from the indica group showed the smallest response (Figure 10). Indica x japonica varieties seem to be dominated by the indica group in their response.

Confirming the characterization by Te-Tzu Chang (8), the javanica varieties in our test showed the tallest plant habitus 19 days after treatment.

Comparative study of Tetcyclacis and other bioregulators. Many growth regulators are known to inhibit cell elongation, whereas only a few have present economic importance on various crops other but rice. Tests in a closed system have shown quaternary ammonium compounds to be less effective compared to the highly active compounds, Tetcyclacis and Ancymidol (4). These findings have been confirmed in trials under conditions comparable to a seedling center (Figure 11).

The most dramatic reduction in plant height could be seen after seedsoaking with methyl-Chlorfluorenol. However, this compound caused serious phytotoxic symptoms. Most promising was Tetcyclacis. This bioregulator showed a plant height reduction which was significantly superior to all other tested compounds. By increasing the application rate from 10 to 50 and 100 ppm Tetcyclacis and Ancymidol showed equal plant height reductions 19 days after seeding.

All the other substances, except methyl-Chlorflurenol which was highly phytotoxic, had no statistically significant effect (Figure 12).

Comparing the different substances 19 days after seeding by their effect on length inhibition on the 2nd leaf sheath, at the 10 ppm rate, Tetcyclacis is significantly superior to all other compounds (Figure 13).

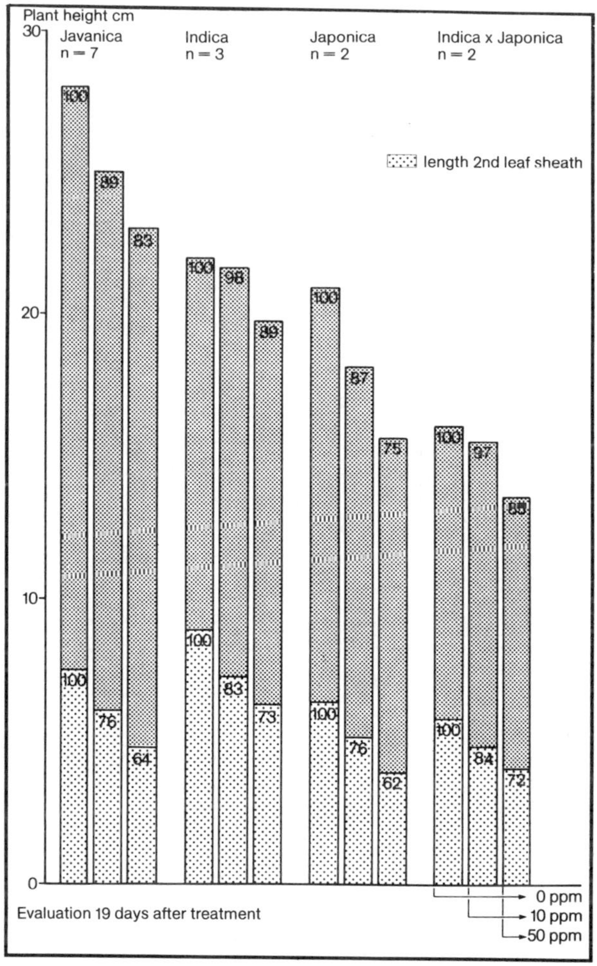

Figure 10 Influence of a seedsoaking treatment with Tetcyclacis on plant height and length of 2nd leaf sheath of Oryza sativa from various origins

10 ppm	plant height					
	days after treatment (seeding)					
	7 (4)		15 (12)		22 (19)	
Untreated	3,3	D	16,4	D E F	19,5	C
Tetcyclacis	0,9	A	8,0	B	12,9	A B
methyl-Chlorflurenol	1,5	B	5,8	A	12,2	A
Ancymidol	2,2	C	10,8	C	16,2	A B C
Lab 117 682 W	3,0	D	15,6	D E F	18,0	B C
Chlormequat chloride	3,2	D	14,8	D	17,7	B C
Daminozide	3,3	D	15,4	D E	18,9	C
Ethephon	3,2	D	15,8	D E F	19,8	C
Mefluidide	3,1	D	16,3	D E F	20,1	C
Maleic hydrazide	3,3	D	17,2	F	20,0	C
Mepiquat chloride	3,3	D	16,9	E F	21,3	C
LSD 95 % Tuckey-test different letters differ significantly	**0,42**		**1,64**		**5,13**	

Figure 11 Comparative study of Tetcyclacis and other bioregulators - Relation between various regulators and the height of Oryza sativa, cv. Koshihikari until transplanting -

ppm	0	10	50	100	Tuckey test LSD 95%
methyl-Chlorflurenol	*DE	AB	A	A	5,13
Tetcyclacis		ABC	AB	A	
Ancymidol		BCD	A	A	
LAB 117 682 W		CDE	DE	CDE	
Mefluidide		DE	DE	DE	
Maleic hydrazide		DE	DE	DE	
Ethephon		DE	DE	DE	
Mepiquat chloride		DE	E	E	

*different letters differ significantly

Figure 12 Influence of various bioregulators on the plant height of Oryza sativa, cv. Koshihikari (japonica) seedlings 19 days after seeding

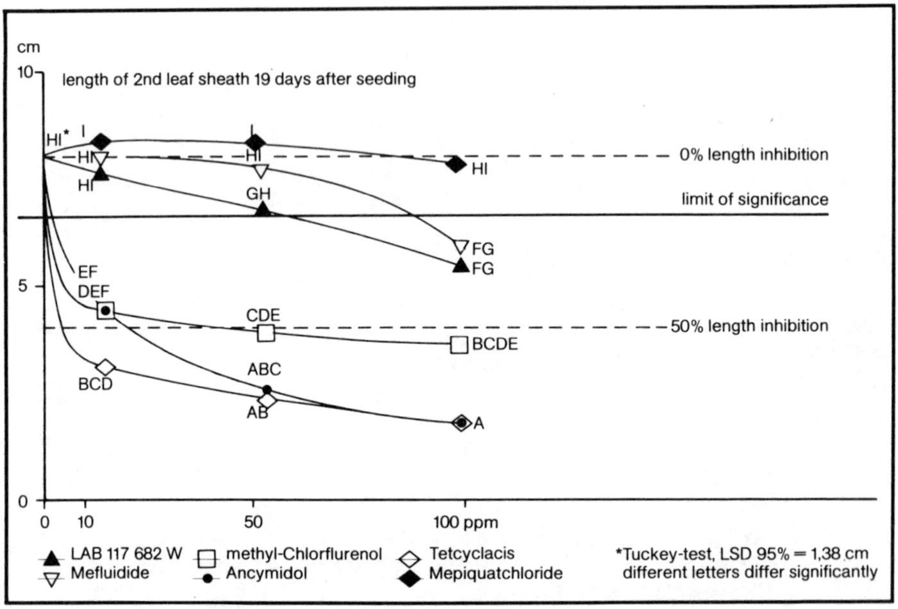

Figure 13 Influence of various bioregulators on the length of 2nd leaf sheath of Oryza sativa, cv. Koshihikari (japonica) after seedsoaking treatment

By increasing the application rate to 50 or 100 ppm Tetcyclacis and Ancymidol are equal and still significantly superior to all other compounds tested.

It is surprising that quarternary ammonium compounds like Chlormequat chloride and Mepiquat chloride are ineffective on rice. Both substances are widely used in cereals for shortening and strengthening the haulms to increase lodging resistance. On the other hand, it is also known that small rates of Mepiquat chloride stimulate biomass production in certain crops (6).

Influence of seedsoaking treatment on yield. The new bioregulator Tetcyclacis has been proven to be suitable for programming rice seedling propagation. It is possible to delay transplanting by approximately one week. The treated seedlings are compact and show increased rooting and leaf formation. The transplantation stress of such treated seedlings is decreased and the total growth after transplantation is promoted. The possible influence on the yield was therefore of interest.

Eight results from Taiwan and two results from Japan showed no significant effect on the yield (Figure 14).

It should be noted however, that the high rate of 10 ppm was tested in the cold February growing season in Taiwan. From other investigations it is known, that under cold growing conditions, the application rate has to be lower than 10 ppm. Using the high rate of 10 ppm under cold growing conditions leads to an undesired strong growth retardation. The positive results of a lower rate have been proven by using the application rate of 6 ppm under cold conditions.

Acknowledgments

The authors thank Mr. H. Rosebrock of the BASF Agricultural Research Farm, Ebina, Tokyo, Japan, and Mr. H. Meumann of the BASF Agricultural Research Station Limburgerhof, Germany, for their diligent support in carrying out and assessing the trials described in this paper.

	Tetcyclacis ppm					Tuckey LSD 95%	Number of trials
	0	5,0	6,0	7,5	10,0		
yield dt/ha	45,5	±0	–	–	–	n. s.	1 (Japan)
%	100	100	–	–	–		
dt/ha	84,0	–	+ 3,6	–	–	n. s.	3 (Taiwan)
%	100	–	104	–	–		
dt/ha	45,5	–	–	+ 1,8	–	n. s.	1 (Japan)
%	100	–	–	104	–		
dt/ha	55,8	–	–	–	– 0,9	n. s.	5 (Taiwan)
%	100	–	–	–	98		

Figure 14 Influence of a seedsoaking treatment on the yield of Oryza sativa

Literature Cited

1. De Datta, S.K. "Principles and Practices of Rice Production"; Wiley - Interscience Publication, John Wiley & Sons: New York, Chichester, Brisbane, Toronto, 1981; pp. 222-228.
2. Hildebrandt, E.; Graebe, J.E.; Rademacher, W.; Jung, J. "Mode of action of new plant growth retardants: BAS 106 .. W and triazole compounds"; 11th Int. Conf. on Plant Growth Substances; Aberystwyth, UK, 1982; Abstract No. 762; p. 65.
3. Jung, J.; Koch, H.; Rieber, N.; Würzer, B. J. Agronomy & Crop Science 1980, 149, 128-136.
4. Rademacher, W.; Jung, J. J. Agronomy & Crop Sciences 1981, 150, 363-371.
5. Rieber, N.; Böhm, H. J. Heterocycl. Chem. 1981, 18, 1-7.
6. Schott, P.E.; Rittig, F.R. in "Chemical Manipulation of Crop Growth and Development"; McLaren, J.S., Ed.; Butterworth Scientific: London, Boston, Durban, Singapore, Sydney, Toronto, Wellington, 1982; pp. 415-424.
7. Tanaka, T.; The Bulletin of the National Inst. of Agricultural Sciences (Japan) 1972, Series A, No. 19, 1-100.
8. Chang, T.T.; Bardenas, E. "The Morphology and Varietal Characteristics of the Rice Plant"; TECHNICAL BULLETIN No. 4, IRRI: Manila, 1965, p. 24.
9. Yoshida, S. "Fundamentals of rice crop sciences"; IRRI: Manila, 1981; pp. 33-35.
10. Zadoks, I.C.; Chang, T.T.; Konzak, C.F. Eucarpia Bulletin 1974, 7, 49-52.

RECEIVED February 22, 1984

6

Chemical Manipulation of Soybean (*Glycine max* L. Merr.) Oil Quality

JUDITH B. ST. JOHN, MERYL N. CHRISTIANSEN, and DANIEL E. TERLIZZI

Beltsville Agricultural Research Center, U.S. Department of Agriculture, Beltsville, MD 20705

Action of the enzyme lipoxygenase (E.C. 1.13.11.12) on linolenic acid causes soybean oil to develop rancidity and reduces oil quality. The formation of linolenic acid and the enzyme lipoxygenase were inhibited. This chapter reports the potential for chemical enhancement of soybean oil quality by treatment of growing plants with pyridazinones.

The substituted pyridazinones are a class of chemicals with multiple actions in higher plants. These actions include: Inhibition of the Hill reaction and CO_2 fixation; inhibition of the formation of chloroplast pigments, ribosomes, and membrane lipids; and alterations in chloroplast ultra-structure (1). These actions may occur singly or in combination, dependent upon the structure of the particular pyridazinone (2, 3). BASF 13 338 [4-chloro-5(dimethylamino)3(2H)-pyridazinone = Sandoz 9785 = BASF 105 00W] is representative of the group of pyridazinones that have the strongest action on membrane lipids. These pyridazinones specifically block the conversion of linoleic (18:2) acid to linolenic (18:3) acid. Sandoz 6706 [4-chloro-5-(dimethylamino)-2-(α,α,α-trifluoro-m-tolyl)-3(2H)-pyridazinone] is representative of the pyridazinones referred to as bleaching agents because of their potent inhibition of chloroplast pigment formation. These pyridazinones also block linolenic acid formation, but are not as effective as the 4-chloro-5-dimethylamino-pyridazin-3-ones. The molecular structure of pyrazon [5-amino-4-chloro- 2-phenyl-3(2H)-pyridazinone] accounts for inhibition of the Hill reaction and photosynthetic CO_2 fixation by the pyridazinones.

Polyunsaturated fatty acids in soybean (Glycine max L. Merr.) seeds contain cis,cis-1,4-pentadiene systems that serve as substrates for the formation of hydroperoxides by the enzyme lipoxygenase (E.C. 1.13.11.12) and by air (autoxidation). Hydroperoxides lead to the development of ranoid oil, greatly reducing oil quality. Our interest was in the potential to improve oil quality by using the pyridazinones to prevent the formation of linolenic acid and/or lipoxygenase activity. The pyridazinones have thus far only been used to prevent linolenic acid formation in vegetative tissues of a number of crop species (4, 5, 6).

Pyridazinone Action on Lipids of Soybean Cotyledons *In Vitro*

Experiments to determine the effects of pyridazinones on linolenic acid production in soybean seeds were carried out by Drs. John Thompson and James Madison at the USDA laboratory in Ithaca, New York, using their previously published procedures (7). Briefly, 0.01 M stock solutions of the test compounds were prepared. An aliquot of this stock solution (80 µl) was then combined with nutrient culture medium (8 ml) to give a final pyridazinone concentration of 0.1 mM. The bottles were capped and sterilized by autoclaving. Pods were harvested from the plants and sterilized with 10-fold diluted Chlorox. The embryos were then removed from the pods aseptically and cut into two relatively equivalent halves (cotyledons). One-half was placed in a bottle containing only the nutrient solution. The other half was placed in a bottle containing the nutrient solution plus the test compound. These bottles were incubated for 6 days at 27°C with slow reciprocal shaking. At the end of the incubation period, the cotyledons were removed from the bottles, washed with water, blotted and weighed. The cotyledons were then lyophilized and lipids were extracted and analyzed by gas chromatography (4).

This split-cotyledon test system eliminates genetic variability between control and treated tissues. In this *in vitro* system BASF 13 338 and Sandoz 6706 reduced the relative proportion of linolenic acid in soybean lipids (Table I). No reduction in cotyledon weight occurred as a result of pyridazinone treatment.

The formation of linolenic acid is generally accepted to occur by elongation of palmitic acid (16:0) to stearic acid (18:0) and subsequent sequential desaturation to oleic acid (18:1), linoleic acid (18:2), and linolenic acid (18:3). The data in Table I show that the relative proportion of linoleic acid increases when linolenic acid decreases, without a change in the overall ratio of saturated to unsaturated fatty acids. Thus the ratio of 18:2/18:3 is a reflection of the inhibition of linoleic acid desaturation to linolenic acid. Conversion of linoleic acid to linolenic acid in soybean cotyledons (seeds) is inhibited by pyridazinones when the cotyledons are in direct contact with the pyridazinones in nutrient cultures.

Table I. Effects of BASF 13 338 and Sandoz 6706 on Relative Proportions of Linolenic Acid in Total Lipids of Soybean Cotyledons

Treatment (0.1 mM)	Linolenic Acid (18:3) % by wt	Ratio of Linoleic Acid (18:2) to Linolenic Acid (18:3)	S/U[1]
Control	14.8	3.41	0.16
BASF 13 338	10.8	5.04	0.15
Sandoz 6706	11.8	4.14	0.15

[1] S/U = palmitic (16:0) + stearic (18:0) acids / oleic (18:1) + linoleic (18:2) + linolenic (18:3) acids.

Pyridazinone Action on Lipids of Soybean Seeds From Plants Grown and Treated in a Greenhouse

Soybeans were grown in Jiffy Mix in 7 inch pots in the greenhouses at USDA, Beltsville, Maryland. Treatments, replicated six times, are indicated in Table II.

Sandoz 6706 applied as a 5 ppm soil drench 3 days after flower induction was the only pyridazinone/treatment combination that resulted in reduced linolenic acid in oil from seeds of greenhouse-grown soybeans. This treatment with BASF 13 338 increased linolenic acid levels. Experiments replicated in time verified the results for the 5 ppm soil drench 3 days after induction of flowering. The percentage of linolenic acid in the oil was measured at 5.7 for the control and 6.3 and 4.8 for BASF 13 338 and Sandoz 6706, respectively. These data are indicative of the complications that can arise in attempts to move from predictable results in an in vitro system (Table I) to the more complex set of variables present in a greenhouse situation (Table II).

Table II. Effects of Pyridazinones on Relative Proportions of Linolenic Acid in Soybean Oil of Seeds from Greenhouse Plants

Treatment	Linolenic acid % by wt. of total lipid
Control	5.0
BASF 13 338	
Soil incorporated, 5ppm	5.0
Spray, 10 ppm 3 days after flower induction	6.0
Soil drench, 5 ppm 3 days after flower induction	6.0
Sandoz 6706	
Soil incorporated, 5 ppm	4.9
Spray, 10 ppm 3 days after flower induction	4.9
Soil drench, 5 ppm 3 days after flower induction	4.4

Effects of BASF 13 338 on In Vivo Lipoxygenase Activity of Soybean Seeds from Field-Grown Plants

All of our experiments to this point had involved attempts to prevent the formation of linolenic acid in the oil of soybean seeds. Dr. Robert Ory, USDA, New Orleans, LA, personally communicated to us his preliminary observation on pyridazinone inhibition of lipoxygenase in peanuts. This communication led us into studies of pyridazinone action on soybean lipoxygenase. We first determined the lipoxygenase activity in soybean seeds from plants grown and treated at the BASF experimental farms in Greenville, Mississippi. In vivo lipoxygenase activity was assayed using soybean homogenates prepared according to Vick and Zimmerman (8) with an 18:2 substrate solution described by Surry (9). Lipoxygenase activity was measured with a YSI oxygraph at 30°C with a Clark O_2-electrode. The data in Table III show that a BASF 13 338 treatment of 2.24 kg ai/ha applied as a foliar spray at flowering reduced lipoxygenase activity to 78 percent of the control level. This 22 percent reduction in lipoxygenase activity was significantly different from the control at the 5% confidence level.

Table III. Effects of BASF 13 338 on *In Vivo* Lipoxygenase Activity of Soybean Seeds From Plants Grown and Treated Under Field Conditions

Treatment	Lipoxygenase Activity % of Control
Pre-emergence, soil incorporated	
4.48 kg ai/ha	100
Post-emergence, broadcast spray	
2.24 kg ai/ha	100
Flowering, foliar spray	
2.24 kg ai/ha	78

We have since obtained data from greenhouse experiments conducted at Beltsville, Maryland, verifying pyridazinone reduction of *in vivo* lipoxygenase activity of soybean seeds (Table IV).

Table IV. Effects of Pyridazinones on *In Vivo* Lipoxygenase Activity of Soybean Seeds From Plants Grown and Treated in Greenhouses

Treatment	Lipoxygenase Activity % of Control
Foliar spray, 50 ppm with 1% ethanol and 0.01% Tween 20	
BASF 13 338	75[1]
Sandoz 6706	70[1]
Norflurazon[2]	76

[1]Significantly different from the control at the 5% confidence level.
[2]Norflurazon = 4-chloro-5-(methylamino)-2-(α,α,α-trifluoro-m-tolyl)-3(2H)-pyridazinone.

All three of the pyridazinones tested significantly reduced lipoxygenase activity in the seeds from soybean plants treated with a foliar spray at flowering. The most promising potential for enhancement of soybean oil quality with the pyridazinones appears to be by reduction of *in vivo* lipoxygenase activity in the seed. Ory et al. in a preliminary report (10) and in the detailed report (11) in this volume have reached similar conclusions based on their work with peanuts.

Effects of Pyridazinones on *In Vitro* Lipoxygenase Activity

The *in vitro* assays were conducted to obtain information on the nature of pyridazinone inhibition of lipoxygenase activity. The *in vitro* assays were carried out as described for the *in vivo* assays except that "crystalline" soybean lipoxygenase was obtained from

Sigma Chemical Company, St. Louis, Missouri. Approximately 30 pyridazinones, including those in Table IV, were evaluated for inhibition in the *in vitro* system. Pyrazon and BASF 37827 [2-(4-fluorophenyl)-4-bromo-5-amino-3(2*H*)-pyridazinone] reduced *in vitro* lipoxygenase activity to 56 and 73 percent, respectively, of the control activity at an inhibitor concentration of 0.1 mM and a substrate concentration of 0.5 mM. The double-reciprocal plot (Fig. 1) of pyrazon action on the "crystalline" soybean lipoxygenase suggests that pyrazon is a competitive inhibitor in this system. The calculated Michaelis-Menton constants for the control were: V max = 0.373 moles/min. K_m = 1.108 mM. The constants for 0.1 mM Pyrazon were: V max = 0.392 moles/min.; K_m = 1.817 mM; K_i = 0.16 mM. Pyrazon and BASF 37827 were not included in our *in vivo* studies. This, combined with the lack of *in vitro* inhibition of "crystalline" soybean lipoxygenase by BASF 13 338, Sandoz 6706 and Norflurazon, suggests caution in extrapolating between the *in vivo* and *in vitro* systems. Thus pyridazinones may competitively inhibit soybean lipoxygenase or, conversely, the action may involve a reduced concentration of the enzyme *in vivo*. Ory et al. (11) have concluded that BASF 105 00W reduced the total lipoxygenase activity in peanuts by reducing the concentration of the enzyme.

The combined data indicate that pyridazinones may enhance soybean oil quality by two mechanisms. Pyridazinones may inhibit the formation of linolenic acid (Tables I and II) and/or decrease the activity of lipoxygenase (Tables III, IV, and Figure 1). Methods and timing of application of the appropriate pyridazinones must be developed so that the enhancement of soybean oil quality is not at the expense of reduced yields.

Other Uses for Pyridazinones as Bioregulators

Sunflowers. Substituted pyridazinones may also have potential as bioregulators to bring about desirable changes in the composition of the fatty acids in sunflower (*Helianthus annuus* L.) seed oil (Table V). The sunflowers were grown and treated by Dr. Marvin Heilman, USDA, Weslaco, TX. BASF 13 338 was sprayed directly on the sunflower seed heads at the first bud stage. Fatty acid analyses were performed in Beltsville, Maryland.

Table V. Effect of BASF 13 338 on Fatty Acid Composition of Sunflower Seed Lipids

	Fatty acid composition (% by wt.)				
Treatment	16:0[1]	18:0	18:1	18:2	18:1/18:2
Control	5.62	4.78	41.05	48.58	0.85
BASF 13 338					
50 ppm spray, 1st buds	5.83	4.55	36.50[2]	53.10[2]	0.69[2]

[1]16:0 = palmitic acid. 18:0 = stearic acid. 18:1 = oleic acid. 18:2 = linoleic acid.
[2]Significantly different from the control at the 5% confidence level.

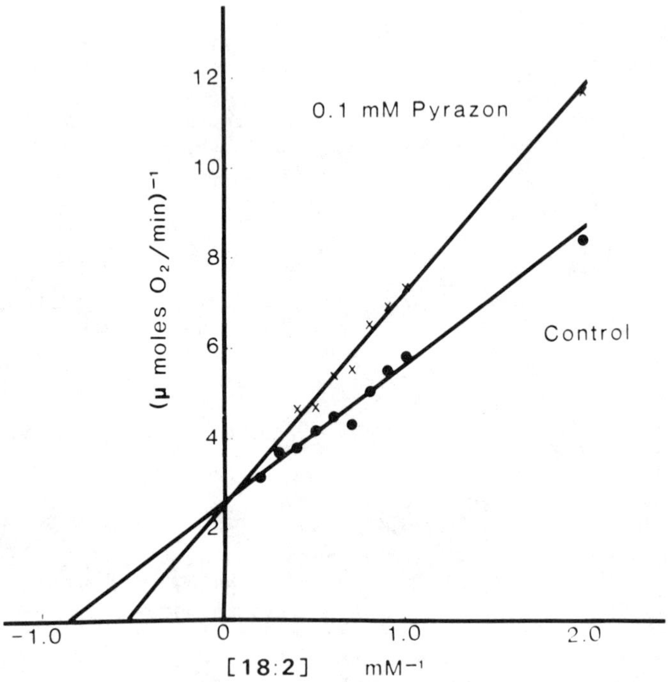

Figure 1. Double-reciprocal plot of lipoxygenase inhibition by pyrazon.

Sunflower oil from seeds produced in cool northern climates normally contain 70% or more linoleic acid, while oil from seeds produced in warmer southern climates may contain as little as 30% linoleic acid. The data in Table V suggest that pyridazinones might be useful in producing sunflower oils with higher levels of linoleic acid. This could be particularly useful in warmer climates and suggests an alternative approach to selective breeding for high-linoleic acid sunflower lines. It is interesting to note that BASF 13 338 increased levels of linoleic acid in a seed that does not contain linolenic acid.

Boll Weevils. The activity of pyridazinones in biological systems other than plants further justifies the term bioregulator as the descripter for their action. The possibility of using a single bioregulator to enhance the quality of a plant product and bioregulate the life cycle of a pest insect is intriguing. Biological activity of the pyridazinones against boll weevils (*Anthonomus grandis*) is documented in Table VI. These experiments were conducted at the USDA Boll Weevil Research Laboratory, Mississippi State, Mississippi.

Table VI. Effects of Feeding Two Substituted Pyridazinones to Boll Weevils (*Anthonomus grandis*)

	BASF 13 338 (% wt/wt added to diet)[1]			
Parameter Measured	Control	0.01	0.03	0.075
Yield, adults/dish	67.8a[6]	64.2a	59.8a	50.0a
Yield, % of control[2]	100a	93.4a	82.3ab	72.8b
Wt. of adults (mg)	13.23a	12.80a	12.01b	10.47c
Eggs/♀/day[3]	8.18a	7.19a	5.07ab	3.52b
Eggs, % of control[4]	100a	91.3a	66.0b	43.1c
Hatch, %[5]	86.4a	81.3ab	78.3ab	69.6b
Viable eggs/day	7.2a	5.9ab	4.0ab	2.6b
	Sandoz 6706 (% wt/wt added to diet)[1]			
	Control	0.01	0.03	0.075
Yield, adults/dish	76.5a[6]	53.3b	44.7b	44.6b
Yield, % of control[2]	100a	70.6b	59.2b	59.2b
Wt. of adults (mg)	13.07a	12.71a	11.66b	10.74b
Eggs/♀/day[3]	6.60a	6.62a	4.77ab	2.79b
Eggs/% of control[4]	100a	100.3a	72.3b	41.3b
Hatch, %[5]	74.2a	65.9a	59.6a	3.04a
Viable eggs/day	4.95a	4.46a	3.04a	1.94a

[1]Larval + adult diets.
[2]Used to eliminate variation in number of eggs and hatch from rep to rep in time.
[3]12 consecutive days (50 pair/rep).
[4]Eliminated variation in time.
[5]Five samples; 300 eggs/sample/replicate.
[6]Means followed by the same letter are not significantly different.

Animal Systems. Lipoxygenase is found in animals as well as plants. Arachidonic (20:4) acid serves as a substrate for lipoxygenase in animal systems. Products of lipoxygenase action on arachidonic acid have been implicated in allergic and inflammatory reactions in mammals (12). Pyrazon was a more effective inhibitor of soybean lipoxygenase activity when arachidonic acid was used as the substrate (Table VII).

Soybean lipoxygenase is the only commercially available form of the enzyme. It has been used as a model system for studying inhibitors of in vivo mammalian lipoxygenase (13, 14). Wallach and Brown (14) concluded that the response to inhibitors of soybean lipoxygenase and human platelet lipoxygenase were qualitatively similar. Our results therefore suggest a potential pharmacological role for the pyridazinones as lipoxygenase inhibitors.

Table VII. Effect of Pyrazon on In Vitro Lipoxygenase of "Crystalline" Commercial Soybean Enzyme

Treatment (0.1 mM)	Substrate (0.5 mMO)	% of Control[1]
Pyrazon	Linoleic acid	62
Pyrazon	Arachidonic acid	36

[1]Control = 0.125 moles O_2/min. with linoleic acid as substrate and 0.177 moles O_2/min. with arachidonic acid as substrate.

Summary. The substituted pyridazinones are bioregulators with the potential for numerous applications in agriculture. The quality of soybean products may be improved through the reduction of undesirable compounds (linolenic acid) and reduction of undesirable enzymic activity (lipoxygenase). The pyridazinones offer the potential for chemically enhancing linoleic acid levels in sunflowers. Additionally, the possible pharmacological role in animal systems for the pyridazinones suggests the desirability of developing chemicals with broad applications.

Acknowledgments

This work was supported in part by competitive grant #80433 from the American Soybean Association to Judith B. St. John and Meryl N. Christiansen.

Literature Cited

1. Eder, F.A. Z. Naturforsch, 1979, 34C, 1052-54.
2. St. John, J.B. in "Biochemical Responses Induced by Herbicides;" Moreland, D.E., St. John, J.B., and Hess, F.D., Ed.; ACS SYMPOSIUM SERIES NO. 2, American Chemical Society: Atlanta, Georgia, 1981; pp. 97-110.
3. St. John, J.B.; Schirmer, U.; Rittig, F.R.; and Bleiholder, H. in "Rational Approaches to Pesticide Synthesis;" Kohn, G.K., and Menn, J.J., Ed.; ACS SYMPOSIUM SERIES, American Chemical Society: Washington, D.C., 1983; In Press.
4. St. John, J.B.; Christiansen, M.N. Plant Physiol., 1976, 57, 257-59.
5. St. John, J.B.; Christiansen, M.N.; Ashworth, E.N.; Gentner, W.A. Crop Sci., 1979, 19, 65-9.
6. Willemot, C.; Slack, C.R.; Browse, J.; and Roughan, P.G. Plant Physiol., 1982, 70, 78-81.

7. Thompson, J.F.; Madison, J.T.; Muenster, A.E. Ann. Bot., 1977, 41, 29-39.
8. Vick, B.A.; Zimmerman, D.C. Plant Physiol., 1976, 57, 780-788.
9. Surry, K. Plant Physiol., 1964, 39, 65-70.
10. Ory, R.L.; Conkerton, E.J.; St. Angelo, A.J.; Vincent, C.H.; Rittig, F.R.; Schroeder, M. Proc., Amer. Peanut Res. Educ. Soc., 1981, 13, 86.
11. Ory, R.L.; St. Angelo, A.J.; Conkerton, E.J.; Chapital, D.C.; and Rittig, F.R. in "Bioregulators: Chemistry and Uses;" Ory, R.L., Rittig, F.R., Ed.; ACS SYMPOSIUM SERIES, American Chemical Society, Washington, D.C., 1983; this book.
12. Marx, J.L. Science, 1982, 215, 1380-3.
13. Blackwell, G.J.; Fowler, R.J. Prostaglandins, 1978, 16, 417-25.
14. Wallach, D.P.; Brown, V.R. Biochim. Biophys. Acta, 1981, 663.

RECEIVED February 22, 1984

7

Trigonelline and Promotion of Cell Arrest in G2 of Various Legumes

LANCE S. EVANS and WILLIAM A. TRAMONTANO

Laboratory of Plant Morphogenesis, Manhattan College, Bronx, NY 10471

Trigonelline, present in dry seeds of many legumes is transported to enlarging roots and shoots during early seedling ontogeny. This hormone promotes cell arrest in G2 in 20-40% of all root cells during normal cell differentiation in roots of *Pisum sativum*. In the absence of trigonelline this cell population arrests in G1. Results presented herein show that trigonelline also promotes cell arrest in G2 in roots of *Glycine max* and *Phaseolus vulgaris* and that the percentage that arrested in G2 in roots of *G. max* decreases during seedling ontogeny as it does in *P. sativum*. During development, trigonelline is synthesized in leaves and is translocated to pods and eventually to seeds during fruit maturation in *P. sativum* and *G. max*. Seeds of most legumes had high concentrations of trigonelline and those of some non-legumes had low concentrations. These results demonstrate that trigonelline is present in several legumes and promotes cell arrest in G2 in these species.

0097–6156/84/0257–0075$06.00/0

Trigonelline functions as a plant hormone in Pisum sativum [1-3]. Trigonelline, found in cotyledons of dry seeds is transported from cotyledons to other plant tissues during early seedling development and promotes preferential cell arrest in G2 of the cell cycle.

Arrest in the cell cycle of meristematic cells in cultured primary root tips after temporary carbohydrate deprivation (to produce stationary phase meristems) is a non-random process [4]. Cell arrest during normal cell differentiation (e.g., in more mature root segment segments) also occurs as a non-random process. Results have also shown that the proportions of cells with 2C and 4C DNA contents in stationary phase meristems is similar to proportions in more mature root segments [15]. Few cells arrest during DNA synthesis and no cells arrest during mitosis. Under temporary carbohydrate deprivation root cells of Pisum sativum arrest in G1, in G2, or become polyploid. If trigonelline is present, as high as 60% of the meristem cell population arrests in G2. If sufficient trigonelline is not present, these cells arrest only in G1 [2, 4, 6].

Trigonelline has been detected in many plants and animals. In plants, it has been found in several Gymnospermae, in several Monocotyledoneae, and many Dicotyledoneae [7]. Moreover, trigonelline has been found in a wide variety of animals [8]. Many mammals excrete trigonelline in urine [9]. No hormone effects of trigonelline have been found in animals.

Trigonelline is usually the most abundant molecule in the pyridine nucleotide metabolic pathway for the production of nicotinamide adenine dinucleotide (NAD) [10-12]. These results indicate that trigonelline may have a regulatory role in NAD biosynthesis as well as in cell proliferation.

Experiments reported herein were performed to determine:

(1) if trigonelline promotes cell arrest in G2 in other legumes besides P. sativum,
(2) if the proportion of cells arrested in G2 decreases in soybean roots as a function of seedling age as it does in roots of P. sativum,
(3) trigonelline concentrations in various tissues of peas and soybeans during ontogeny, and
(4) concentrations of trigonelline in dry seeds of various plant species.

Materials and Methods

A stationary phase, meristem may be defined as a meristem (0-2 mm portion) in which progression through the cell cycle has ceased temporarily [5]. In some experiments trigonelline (Sigma Chemical Co., St. Louis, MO) was added to the sucrose medium which normally supports growth of ten excised roots. After eventual establishment of a stationary phase by temporary carbohydrate deprivation, cells were arrested either in G1 or in G2 (2C and 4C contents, respectively) within the terminal meristem. For example, if roots from 3-day-old seedlings were placed in sucrose medium alone before establishment of stationary phase, only 0.20 cells arrest in G2. However, if a sufficient concentration of trigonelline was present in medium with sucrose before establishment of stationary phase, a larger proportion (≈ 0.40 - 0.60) of cells arrest in G2 [2]. The stationary phase technique was used to determine if root meristems of G. max and P. vulgaris respond to trigonelline.

DNA measurements. Measurements of relative DNA per nucleus of Feulgen-stained nuclei were obtained by microfluorimetry. This method is modified from that of ref [19] and is described elsewhere [4].

Intact plant studies. Garden pea and soybean plants were grown to maturity in a soil mixture which consisted of 40% topsoil, 40% perlite, 20% peat, buffered with lime. The appropriate Rhizobium inoculum was also added. Plants were sacrificed every 10 days so that various organs and tissues could be extracted to determine trigonelline concentrations.

Trigonelline concentrations. To determine concentrations of trigonelline, all tissues from plants at each age were separated into roots, stems, leaves, pods, and seeds and extracted in an ethanol series [10]. Each extract was concentrated and spotted on Silica gel TLC UV plates, 250 μm thickness (Analtech Corp., Newark, DE, USA). Extracts were developed in acetone-water (1:1). Plates were allowed to air dry and trigonelline was eluted. Quantities of trigonelline were determined by HPLC [20] using a Whatman Partisil-Sax 10 column 7 mM KPi as solvent (pH 5.8) at a flow rate of 1 ml/min.

Results

Since most of the previous research to demonstrate that trigonelline is a hormone that influences cell cycle kinetics of roots and shoots was done with P. sativum, experiments were designed to determine if trigonelline had similar effects in two other legumes. When excised roots of G. max and P. vulgaris (Table 1) were exposed to trigonelline, preferential cell arrest in G2 was evident. A concentration of 10-4 M trigonelline in the bioassay increased the proportion of cells arrested in G2 by 0.14 and 0.12 in G. max and P. vulgaris, respectively.

Since trigonelline promoted preferential cell arrest in G2 of G. max and P. vulgaris in aseptic culture of excised roots, experiments were performed to determine if the proportion of cells arrested in G2 in G. max naturally decreases during seedling ontogeny as it does in P. sativum [3]. When excised roots from 3-, 5-, 7-, and 10-day-old plants of G. max were placed in medium without sucrose for three days to establish a stationary phase, proportions of cells arrested in G2 were 0.50, 0.34, 0.17, and 0.11 (Table 2), respectively. The proportion of cells arrested in G2 decreases during early seedling ontogeny.

Previous results suggested that the total amount of trigonelline in seedlings of P. sativum is relatively constant during the first 7 days after germination and that the amount increased by about 25% in 10-day-old seedlings [13]. Experiments were performed to determine the concentration of this hormone during plant development (Table 3). As seedling age increases trigonelline concentrations decrease in roots and stems to 1 μg/g tissue or lower. From 20 to 60 days the concentration of trigonelline in leaves decreases from 80 to 8.8 μg/g tissue.

Table 1. Demonstration that trigonelline promotes preferential cell arrest in G2 stationary phase meristems of soybeans (*Glycine max*) and pinto beans (*Phaseolus vulgaris*)

Treatment	Proportion of cells in G2	
	Soybeans	Pinto beans
3-day-old seedlings - excised roots placed in medium without carbohydrate for 4 days	0.50±0.08[a)]	0.41±0.04[a)]
3-day-old seedlings - excised roots grown in medium with sucrose without trigonelline before placement in medium without sucrose for 4 days	0.18±0.03	0.18±0.03
3-day-old seedlings - excised roots grown in medium with sucrose and 10^{-4} M trigonelline before placement in medium without sucrose for 4 days	0.32±0.02	0.30±0.03

a) Values represent mean and standard error, respectively.

Table 2. Proportion of cells in G2 stage of the cell cycle in stationary phase meristems of excised roots of soybeans as a function of seedling age

Seedling age (days)	Proportion of cells in G2
3	0.50±0.08
5	0.34±0.04
7	0.17±0.03
10	0.11±0.06

a) Values represent mean and standard error, respectively.

In forty-day-old plants a high trigonelline concentration (203 µg/g tissue) is present in pods but most of this trigonelline is present in seeds as plants mature. At maturity, most of the trigonelline in the entire plant is present in the seeds.

Since roots of G. max respond to trigonelline similar to roots of P. sativum, experiments were performed to determine trigonelline concentrations in various organs of plants of G. max. Concentrations of trigonelline were below 1.5 µg/g tissue in roots throughout ontogeny. Trigonelline concentrations were below 0.5 µg/g in stem tissues from 20 to 60 days. A much higher level (7.6 µg/g tissue) was present in 70-day-old stem tissue but the days. A much concentration decreased thereafter (Table 4). Trigonelline concentrations were relatively high (63 µg/g tissue) in leaves of 20-day-old plants but decreased gradually so that 100-day-old plants had only about 10 µg/g tissue. Trigonelline concentrations in pods of 60- to 70-day-old plants were about 15 µg/g tissue but decreased to about 4 µg/g tissue matured. Seeds have relatively constant concentrations of about 20 µg/g fresh tissue.

These results show that roots of G. max and P. vulgaris exhibited promotion of cell arrest in G2 with added trigonelline similar to that of P. sativum. Moreover, roots of Helianthus annuus and Triticum aestivum do not respond to trigonelline [14]. To determine if trigonelline may have a hormonal role in several other plant species, assays were performed to determine trigonelline concentrations of several plant species (Table 5). Except for Arachis hypogea (peanut) all legumes tested had relatively high trigonelline concentrations. In contrast, most non-legumes and monocots had relatively low trigonelline concentrations in dry seeds.

Discussion

Trigonelline has been shown to promote cell arrest in G2 in roots of four legumes, Pisum sativum, Vicia faba, Glycine max, and Phaseolus vulgaris. It does not seem to influence two non-legumes tested (Triticum aestivum and Helianthus annuus); [14]. Although more species must be tested, it seems that trigonelline affects only legumes by promoting cell arrest in G2. The reason(s) for this apparent specificity is unknown.

Predominant cell arrest in G2 in roots is common within but is not restricted to legumes. Several other non-legume species such as Allium cepa [15], Zea mays [16], and Raphanus sativa [17] exhibit preponderant cell arrest in G2 in mature roots. Additional species, such as these three, should be tested to determine if they respond to trigonelline since both T. aestivum and H. annuus exhibit predominant cell arrest in G1 [5].

Trigonelline concentrations are high in seeds of most of the legumes tested and are markedly lower in some non-legumes. However, it cannot be concluded that high trigonelline concentrations are restricted only to legumes. Seeds of Coffea arabica have large concentrations of trigonelline [18]. C. arabica is in the Rubiaceae and is not distantly related to the Leguminosae. Also, seeds of Secale cereale (rye) contained (52 µg/g tissue) concentrations of trigonelline. Thus, no relationship between high trigonelline concentrations and preponderant cell arrest in G2 only in legumes can be established with available data.

Table 3. Trigonelline concentrations in various tissues of Pisum sativum during ontogeny

Plant age (days)	Trigonelline concentration (μg/g fresh mass)				
	Roots	Shoots	Leaves	Pods	Seeds
20	6.3±0.2	24.5±5.3	88.2±6.0	-	-
30	6.7±3.0	19.8±1.0	8.0±0.0	-	-
40	0.7±0.2	3.5±0.2	11.1±3.0	202.6±4.0 a)	-
50	*b)	1.2±0.5	20.2±0.0	37.7±7.0	127.6±2.2
60	1.0±0.6	*	8.8±4.6	5.8±4.8	166.4±16.0

a) Immature seeds were not separated from the pods at this point because seeds were very small and would have an insignificant mass compared with the mass of the pods. Values represent mean and standard error, respectively.

b) Asterisk denote samples below 0.5 μg/g tissue, which is the lowest limit of detection of trigonelline with the procedure used.

Table 4. Trigonelline concentration in various fresh tissues of Glycine max during ontogeny

Plant age (days)	Trigonelline concentration (μg/g fresh mass)				
	Roots	Stems	Leaves	Pods	Seeds
20	*b)	*	63.2±4.4	-	-
30	*	*	40.3±5.6	-	-
40	*	*	48.8±2.2	-	-
50	*	*	27.8±2.3	-	-
60	*	*	39.4±2.0	16.2±2.0 a)	-
70	*	7.6±1.0	19.5±1.9	14.8±1.4 a)	-
80	1.1±0.05	1.5±1.0	18.3±11.6	5.3±0.4 a)	-
90	*	2.0±1.0	10.8±0.4	4.5±0.5	19.7±2.6
100	*	3.0±0.5	10.5±4.5	3.7±2.0	21.7±3.3

a) Immature seeds were not separated from the pods at this point because seeds were very small and would have an insignificant mass compared with the mass of the pods. Values represent mean and standard error, respectively.

b) Asterisks denote samples below 0.5 μg/g tissue, which is the lowest limit of detection of trigonelline with the procedures used.

Table 5. Trigonelline concentration in dry seeds of various plant species

Plant species	Trigonelline concentration (µg/g tissue)
Dicotyledonous plants	
Legumes	
Pisum sativum	227±11.3
Vicia faba	294±14.7
Glycine max	71.8±5.7
Phaseolus vulgaris (pinto bean)	344±17.2
Phaseolus vulgaris (white bean)	554±13.8
Arachis hypogea (peanut)	8.6±3.3
Lens culinaris (lentil)	363±18.1
Psophocarpus tetragonolobus (winged bean)	384±19.2
Non-legume	
Raphanus sativus (radish)	7.60±2.1
Solanum melongena (eggplant)	1.10±0.0
Cucurbita maxima (squash)	1.80±0.5
Lactuca sativa (lettuce)	11.7±2.0
Capsicum annuum (pepper)	0.60±0.0
Cucumis melo (muskmelon)	2.60±0.0
Monocotyledonous	
Hordeum vulgare (barley)	8.90±4.0
Triticum aestivum (wheat)	*a)
Zea mays (maize)	4.00±2.0
Secale cereale (rye)	52.0±2.6
Sorghum halepense (sorghum)	14.9±4.9
Allium cepa (onion)	13.0±2.0

a) Asterisk denotes sample below 0.5 µg/g tissue, which is the lowest limit of detection of trigonelline with the procedures used. Values represent mean and standard error, respectively.

Trigonelline contents in various plant organs change during ontogeny in the two legumes tested. In general, the concentration in young seedlings is the same as that in dry seeds suggesting that trigonelline is not synthesized and/or destroyed during that time period. After seedling establishment, trigonelline seems to be synthesized in leaves. It is transported to pods and eventually to developing seeds. The reasons why trigonelline is synthesized primarily in leaves and why it is stored in seeds are unknown at present.

The decrease in trigonelline concentration in roots during seedling ontogeny is highly correlated with the decrease in the proportion of cells arrested in G2 [3]. The decrease in trigonelline concentration in roots continues throughout ontogeny. No data are available to determine if the proportion of cells arrested in G2 also decreases in seedlings 10 days after germination begins. Such information is necessary to more fully evaluate trigonelline's role in cell cycle kinetics throughout ontogeny.

At the present time, available information suggests that trigonelline is a hormone that is synthesized in leaves of legumes and is transported to seeds during maturation. During germination this stored trigonelline is retranslocated to enlarging roots and shoots where it is responsible for promoting cell arrest in G2 as cells mature. This hormonal function of trigonelline may occur concurrently with its non-hormone role as a storage form of nicotinic acid for the pyridine nucleotide metabolic cycle for the synthesis of NAD [10, 11]. Additional research is necessary to clarify the various metabolic and hormonal functions of trigonelline.

Literature Cited

1. Lynn, D. G., Nakanishi, K., Patt, L. S., Occolowitz, J. L., Almeida, M. S. and Evans, L. S. (1978) J. Am. Chem. Soc. 100, 7759.
2. Evans, L. S., Almedia, M. S., Lynn, D. G. and Nakanishi, (1979) Science 203, 1122.
3. Evans, L. S. and Tramontano, W. A. (1981) Am. J. Botany 68, 1282.
4. Evans, L. S. (1979) Am. J. Botany 66, 880.
5. Evans, L. S. and Van't Hof, J. (1974) Am. J. Botany 61, 1104.
6. Evans, L. S. and Van't Hof, J. (1973) Exp. Cell Res. 82, 471.
7. Willeke, U., Heeger, V., Meise, M., Newmann, H., Schindelmeiser, I., Vordemfelde, K. and Barz, W. (1979) Phytochemistry 18, 105.
8. Beers, J. R. (1967) Comp. Biochem. Physiol. 21, 11.
9. Handler, P. and Dean, W. J. (1942) J. Biol. Chem. 146, 357.
10. Godavari, H. R. and Waygood, E. R. (1973) Can. J. Botany 48, 2267.
11. Preiss, J. and Handler, P. (1957) J. Am. Chem. Soc. 79, 4246.
12. Tramontano, W. A., Hartnett, C. M., Lynn, D. G. and Evans, L. S. (1982) Phytochemistry 21, 1201.
13. Tramontano, W. A., Lynn, D. G., and Evans, L. S. (1983)

RECEIVED February 22, 1984

8

Properties of Peanuts (*Arachis hypogaea* L.) from Bioregulator-Treated Plants

ROBERT L. ORY, A. J. ST. ANGELO, E. J. CONKERTON, and D. C. CHAPITAL

Southern Regional Research Center, Agricultural Research Service, U.S. Department of Agriculture, New Orleans, LA 70179

FALK R. RITTIG

BASF Aktiengesellschaft, Landwirtschaftliche Versuchsstation, D-6703 Limburgerhof, Federal Republic of Germany

Peanuts from untreated plants and plants treated with a bioregulator (BR) were harvested, dried, hand-shelled, and analyzed for protein profiles, free sugars, several minerals, lipid peroxide development, and lipoxygenase activity. When applied to the plants at flowering, there were little or no changes in minerals, proteins, or carbohydrates, but there was a significant decrease in lipoxygenase activity of treated peanuts. For another BR, roots and stems of the plants were also collected and dried for mineral analyses. Peanut hulls, skins, roots, stems, and seeds showed no major changes in Ca, Fe, Zn, Cu, and Se levels. The implications of decreased lipoxygenase activity on flavor and shelf life of raw peanuts is discussed.

The growth in world population is increasing our need for more sources of food, particularly protein and low cost nutritious plant foods. It has been estimated that by the year 2020, world population will almost double (1,2,3). World-wide production of food will have to increase by almost as much as it has since the dawn of recorded history and with much less land than was available to our forefathers (1,3-6). Cereal grains, legumes and oilseeds are the major crops expected to meet these needs but new methods must be developed to increase yields, maintain quality, and eliminate post-harvest losses during storage and transportation. Elimination of post-harvest losses could almost double current food supplies in some countries. Plant breeding is presently the ideal method for increasing yields, nutritional quality, resistance to spoilage, diseases, insects or environmental stress in crops, but breeding can take 10-20 years to achieve these goals. Bioregulators (plant growth regulators) are relatively new chemical agents that, in very small amounts, affect specific physiological processes of plants and can achieve desirable changes within one crop year. In 1975, less than 5% of agricultural chemicals sales consisted of plant growth regulators (7) but the future potential for these compounds is very

0097-6156/84/0257-0083$06.00/0

promising (1,4,8,9). Bioregulators are currently at the same stage of development and acceptance as herbicides were 25-30 years ago.

Peanuts (groundnuts; Arachis hypogaea) are one of the major oilseeds grown in tropical and subtropical countries. World production of peanuts in 1983 was estimated at 1.84 million tons (10). India and China, the largest peanut-producing countries, grow them primarily for the oil (11). The United States produces only 10% of the world crop but consumes most of this directly; half in the form of peanut butter and another 30-35% as roasted nuts, in candies, cookies, or other confections. In countries that crush peanuts for the oil, the meal (55-60% protein) is fed to animals or used as fertilizer. Any other use of this meal as a source of food protein is minimal, primarily because of aflatoxin contamination, crude methods for expressing the oil, and/or oxidative rancidity problems. Peanut oil does not contain linolenic acid, a source of lipid peroxides in vegetable oils, but peanuts contain linoleic acid (substrate) and lipoxygenase, the principal catalyst of oxidative rancidity in oilseeds (12,13).

Most investigators agree that bioregulators show promise for increasing yields of peanuts but the results obtained to date are variable (8). Yield increases as a result of bioreguator treatment have been reported in Israel (14) and Sierra Leone (8) but there are other reports of either no effect or only slight increases (8). None of the papers on use of bioregulators on peanuts cited by Nickell (8) reports effects of bioregulators on biochemical or physiological factors that affect stability, shelf life, flavor or nutritional quality of peanuts. This paper reports the effects of bioregulators on biochemical factors and nutritional components, selected minerals, and enzymes that affect shelf life and flavor of stored peanuts.

Bioregulator Treatments. All peanuts were grown in field test plots by Drs. M. Schroeder and T. O. Ware at the BASF Agricultural Research Station in Greenville, Miss. Three bioregulators were tested for their effects on various components and quality factors in peanuts: 1,1-dimethyl-piperidinium chloride (PIX), 4- chloro-5-(dimethylamino)-2-phenyl-3(2H)-pyridazinone (BAS 105 00 W; also called BAS 13 338 and Sandoz 9785), products of BASF, Limburgerhof, Federal Republic of Germany, and Kylar, product of Uniroyal Chemicals, Bethany, Conn., USA.

Treatments with PIX and Kylar. Peanut plants grown in the field were treated with 0.025, 0.050, or 0.075 kg/ha of PIX at 4 weeks after planting, at bud formation, at bloom, at pegging, and at 2 weeks after pegging. There were 4 replicates/treatment in randomized plots. Untreated plants served as one control and Kylar (0.85 kg/ha at pegging) served as a second control, since it has been used on peanuts for several years to increase yields and reduce vine growth. At harvest the entire plants were dried in windrows and saved for mineral analyses by x-ray diffraction analysis. Peanuts were hand-shelled to remove adhering soil, avoid damage to kernels, and minimize possible activation of lipoxygenase activity on the oil by mechanical shelling.

Effects on Selected Minerals in Peanut Tissues. PIX , an approved bioregulator for use on cotton, almost doubles calcium contents of stems in field-grown cotton plants. It was of interest, therefore, to determine the effects on calcium uptake and some other minerals essential for optimum growth and nutrition of peanuts. Calcium is essential for optimum yields and seed germination (15), for proper fruiting of plants (16), for prevention of peg- and pod-rot (17,18), for hardening shells and improving general peanut quality (19). Hardening of shells is essential because weak shells permit invasion by fungal hyphae of *Aspergillus flavus* and *A. parasiticus* and, with moisture contents of 35% at this stage, aflatoxin production can increase, reducing edible quality of the whole peanuts or peanut meal.

There were no significant changes in calcium uptake in any peanut tissues from treated plants compared to untreated, based upon x-ray diffraction analyses but the amounts in various tissues were different: 0.55% in defatted peanut meal, 0.49% in hulls, 0.89% in stems, and 0.40% in roots. Other minerals that are components of some metalloenzymes in peanuts were assayed by x-ray diffraction analysis: iron (peroxidase, involved in lignification of hulls), copper (polyphenol oxidase, also in lignification), and zinc (alcohol dehydrogenase, involved in flavor changes), because of their potential effects on peanut quality and *Aspergillus* invasion. These minerals showed no changes due to PIX treatment but like calcium, the amounts in each tissue also varied. Kylar-treated plants also showed no changes in mineral contents of treated *versus* untreated peanuts.

Effects of BAS 105 00 W. Substituted pyridazinones have been used as bioregulators in crop protection for several years to manipulate temperature response of plants and to induce specific decreases in linolenic acid content of seeds and tissue membranes of wheat (20-23), in leaf tissue of *Vicia faba* (24), oat stems (25), leaf tissue of spinach, peas, linseed, and wheat (26), and soybeans (27). As linolenic acid decreased, corresponding increases in linoleic acid occurred but no changes were noted in oleic, palmitic, or stearic acids, suggesting that the specific effect of pyridazinones is to block synthesis of linolenic acid. Control of linolenic biosynthesis may vary depending upon plant species and even tissue (22). Peanut storage oil does not contain linolenic acid but it is present in membrane lipids. It was therefore of interest to determine if BAS 105 00 W could affect lipid metabolism and/or some associated enzymes in peanuts. Inhibiition of photosynthesis by pyridazinones in certain plants produced lighter green leaves (22, 24, 26), but this is believed to be an effect of interference with the accumulation of chloroplast pigments (27) rather than effects on chloroplast lipids.

Spanish peanut plants were treated with varying concentrations of BAS 105 00 W at different times as indicated in Table I. Yields were increased slightly when applied at flowering, but the differences between treated and untreated plants were not significant.

Nevertheless, the reported effects on lipid metabolism in other crops led us to subject the peanuts to an extensive series of analyses for effects on product quality. Half of the samples were placed in sealed glass jars and stored at 4°C and at room temperature (24-

Table I. Effects of BAS 105 00 W Treatment on Yields of Spanish Peanuts

Sample	Application Rate lb/acre	Application Time	Yield lb/acre
Control			3602
PN-1[a]	0.05	at flowering	3840
PN-2	0.45	" "	3850
PN-3	0.90	" "	3829
PN-4	0.05	at pegging	3816
PN-5	0.45	" "	3383
PN-6	0.90	" "	3546
PN-7	0.45 + 0.45	at flowering and pegging	3715
PN-8	0.90 + 0.90	" " " "	3651

[a] PN-1, etc., designate sample numbers for the different treatments.

27°C) before the chemical analyses and taste panel evaluation. Portions of the remaining peanuts were hand-shelled, blanched to remove testae, and extracted with hexane to separate the storage oil. Yields of oil and meal were 49-51% for all samples. Protein contents (%N x 5.46) were slightly higher than controls at low treatment levels at flowering (49.3% for controls, 53.6% for PN-1; 51.6% for PN-2) but showed no change from the controls at pegging. The defatted meals were subsequently extracted with 10% NaCl, dialyzed against distilled water, and freeze-dried to prepare protein isolates for analysis by disc gel electrophoresis and immunoelectrophoresis. Only disc gel electrophoresis appeared to show any differences between treated and untreated samples. There were slight differences in banding patterns for minor, faster-moving proteins smaller than arachin, the major globulin in peanuts. Whether these small differences are related to the 2-4% increase in protein content of peanuts treated at flowering is a question to be answered by further analyses.

Legumes are acknowledged as sources of flatulent oligosaccharides, but the amounts in peanuts are lower than in other legumes. Oil-free meal was extracted with 80% ethanol to isolate and separate free amino acids from free sugars as described earlier (28). Sugars were analyzed by high pressure liquid chromatography for quantitation. Results in Table II show that all treatments seem to reduce the amounts of oligosaccharides except possibly the highest application level at both flowering and pegging. Compared to a total sugars content of 12.97% in untreated peanuts, the amounts were reduced to 9.7, 8.5 and 7.57% as the rate of treatment applied at flowering increased. Amounts of all sugars were reduced so that the relative amount of sucrose compared to total sugars remained the same for these three samples (PN-1, PN-2 and PN-3). There was no consistent trend or relationship between sugar contents and treatments at pegging or at both flowering and pegging. Total sugars of peanuts treated at both flowering and pegging (PN-7 and PN-8) were 8.59% at the lower application level and 12.56% at the higher, but the ratio of sucrose to total sugars in these samples was lower - 65% compared to 73% in all of the others. This was accompanied by increases in the fructose and glucose contents. Variations of this type would

Table II. Effects of BAS 105 00 W on Free Sugars in Peanuts

Sample[a]	Stachyose %	Raffinose %	Sucrose %	Glucose %	Fructose %	Total Sugars %
Control	1.55	0.55	9.39	0.90	0.58	12.97
PN-1	1.38	0.43	7.17	0.44	0.30	9.72
PN-2	1.26	0.27	6.33	0.47	0.19	8.52
PN-3	1.06	0.39	5.60	0.36	0.16	7.57
PN-4	1.33	0.46	6.93	0.51	0.36	9.59
PN-5	0.90	0.29	4.87	0.44	0.32	6.82
PN-6	1.34	0.42	6.89	0.46	0.42	9.53
PN-7	1.19	0.42	5.62	0.75	0.61	8.59
PN-8	1.34	0.71	8.33	1.04	0.78	12.56

[a] Samples and treatments are same as in Table I.

not be expected within the same variety of peanuts grown under similar conditions, though amounts have been shown to vary between different varieties (29).

Mineral contents of deoiled peanut meals and calcium contents of hulls were also determined on peanuts from treated and untreated plants. BAS 105 00 W, as with PIX and Kylar, had no effect on Ca content of hulls and seeds, nor on Mg, Fe, Cu, Mn, and Zn levels of seeds.

Effects on Lipoxygenase and Lipid Peroxidation. After harvest, peanuts may be stored a year or more until processed into edible products. The types of handling and storage conditions could activate the lipoxygenase in stored seeds to produce lipid peroxides, lowering quality of the peanuts. Peanut oil does not contain linolenic acid, therefore, the primary effect of pyridazinones on formation of this fatty acid in soybeans noted by others could not be tested. The effects on other aspects of lipid metabolism, such as lipid peroxide content, fatty acid profiles, tocopherol content, lipoxygenase activity, and effects on taste panel character notes (beany and rancid) related to lipid oxidation, were investigated before and after the year's storage.

Fatty acid profiles in the neutral oil extracted with hexane from treated and untreated peanuts were unaffected by BAS 105 00 W. The ratio of oleic/linoleic acids remained constant. Lipid peroxides can be quantitated by spectrophotometric analysis of the conjugated diene hydroperoxides in the oil at 234 nm (12,13) and can serve as an indicator of lipoxygenase activity in raw peanuts (Table III). Results show decreases in lipid peroxides in freshly harvested seeds from treated plants compared to controls. The amounts of lipid peroxides were still generally low after 2 years, but the relative decreases previously seen in treated samples of fresh peanuts were no longer present except in peanut samples PN-2 and PN-4.

The pronounced decrease in measurable lipoxygenase activity of fresh peanuts was the most notable benefit of BAS 105 00 W applications. In a preliminary study on peanuts from plants treated with BAS 105 00 W (1 control and 2 treated), lipoxygenase activity in treated seeds was 30-40% lower than that in controls (30). This first series of analyses of spanish peanuts showed decreased lipoxygenase activity in fresh peanuts. When relative activities were determined on peanuts stored a year at room temperature and at 4°C (Figure 1), lipoxygenase activity was still lower in treated

Table III. Development of Lipid Peroxides in Peanuts after 2 Years in Cold Storage. (Absorbance at 234 nm; oil extracted from 1 g peanuts in 30 ml hexane)

	A234/g/30 ml	
Sample[a]	Fresh	2 Years
Control	0.721	1.240
PN-1	0.705	1.677
PN-2	0.528	0.821
PN-3	0.672	1.219
PN-4	0.586	0.811
PN-5	0.566	1.020
PN-6	0.529	1.077
PN-7	0.530	1.298
PN-8	0.551	1.177

[a] Samples and treatments are same as in Table I.

compared to untreated seeds, but storage temperature also appeared to affect total activity. The prolonged exposure to low temperature also lowered total enzyme activity of control (untreated) peanuts stored at the lower temperature (compare samples RPN-C-2 and CPN-C-2). Cold temperature storage combined with BAS 105 00 W resulted in the lowest total enzyme activity.

Michaelis-Menton constants were calculated on numerous samples of peanuts at several concentrations of enzyme to determine if the change in activity of treated seeds affected affinity of enzyme for substrate or the order of reaction for the enzyme. The Lineweaver-Burke curves appeared to be almost parallel, as illustrated by results of a typical test shown in Figure 2. The affinity of enzyme (lipoxygenase) for substrate (linoleic acid) is virtually unchanged in peanuts from treated plants. This suggests that the enzyme itself is unchanged except for a decreased concentration in seeds from treated plants. Since it was possible that the measured decrease in lipoxygenase activity might be due to increased synthesis of tocopherols (natural antioxidants), tocopherol contents of seeds (extracted with hexane) were measured spectrophotometrically at 295 nm with α-tocopherol acetate as standard. The results showed no significant differences in fresh or stored seeds which could account for the changes in lipoxygenase activity and lipid peroxides in the oil. The primary effect of BAS 105 00 W on lipid metabolism of peanuts apparently is on lipoxygenase.

A portion of peanuts from each series was roasted and evaluated for changes in flavor profile by a trained taste panel of 18 members soon after receipt and again after a year's storage. The effect of reduced lipoxygenase activity on flavor was evident in results of taste panel evaluations. Panelists rated the peanuts for intensity of a total of ten flavor notes, two which are related to lipoxygenase

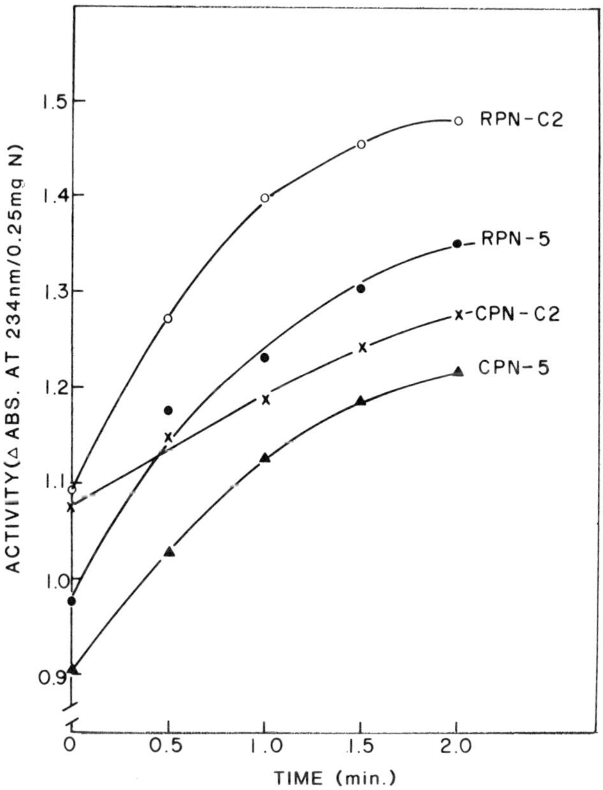

Figure 1. Lipoxygenase activity in treated/untreated peanuts stored a year at two temperatures. RPN-C-2 and RPN-5: Control (untreated) and treated peanuts, respectively, stored at room temp, 24-27 °C; CPN-C-2 and CPN-5: Control and treated peanuts stored at 4 °C. Activity measured at 234 nm as conjugated diene hydroperoxide formed by peanut enzyme (2 mg protein N) from linoleic acid substrate in 3 ml total volume.

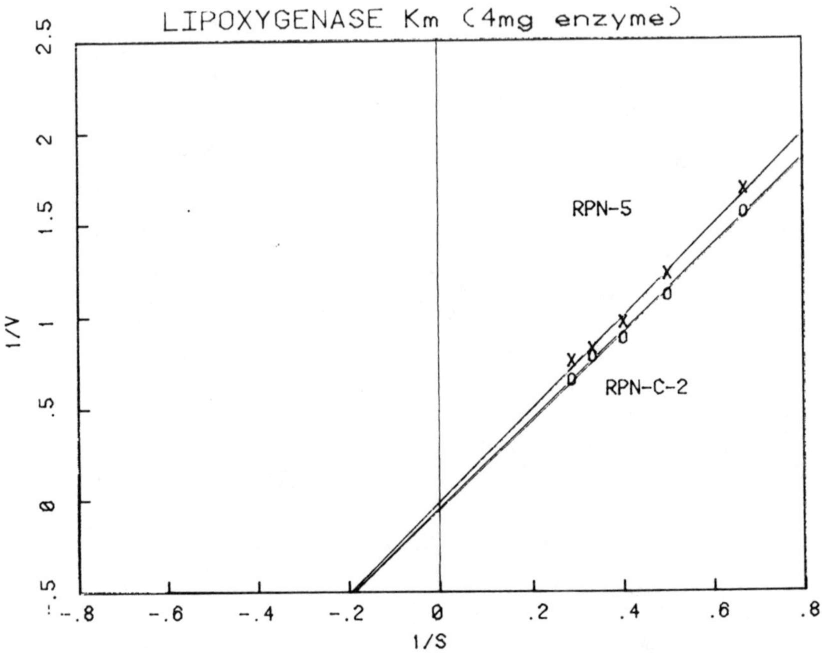

Figure 2. Lineweaver-Burke plots for lipoxygenase activity in treated/untreated peanuts stored a year at room temperature. Conditions: same as for Figure 1. Apparent Km's were: 5.43 x 10^{-5} M for RPN-C-2 and 5.21 x 10^{-5} M for RPN-5.

activity: beany flavor and rancidity. For six parameters and for overall quality of peanuts, differences were not significant, but on an intensity scale of 0 (not detectable) to 9 (extra strong), there were greater differences in beany flavor and rancidity values between treated and untreated peanuts. Ratings for treated peanuts of 4-5 for peanutty flavor (vs. 4 for untreated) and 3.7-5.4 for roasted flavor (vs. 3.7 for untreated) indicate that these were good flavored peanuts, but there was no definite trend based on time of treatment. For example, the lowest and highest treatments at flowering (see Table I) were better than the intermediate level and beany and rancidity characteristics were consistently reduced by all treatments. Ratings of 1.43-2.27 for beany flavor of treated peanuts (vs. 3.25 for untreated) and 0.95-2.06 for rancidity perception (vs. 1.97 for untreated peanuts) suggest that the reduction of lipoxygenase activity in peanuts does translate into improved flavor of roasted peanuts. The potential of BAS 105 00 W as a bioregulator for improvement of flavor and stability of peanuts by reducing lipoxygenase activity is promising.

Acknowledgments

Taste Panel data were obtained by C. H. Vinnett. Technical assistance was provided by T. Scimeca.

Literature Cited

1. Krieger, J. H. Chem. & Eng. News 1982, 60 (No. 51), 9-23.
2. Zikakis, J. P. J. Agric. Food Chem. 1983, 31, 672-675.
3. Kinney, Jr., T. B. Middle Atlantic Regional A.C.S. Meeting, White Haven, PA. April 6-8, 1983, Abst. No. 1.
4. Hardy, R. W. F. ibid. Abst. No. 2.
5. Paulino, L. A. ibid. Abst. No. 5.
6. Marvel, J. T. ibid. Abst. No. 6.
7. Anon. Farm Chemicals 1975, 138 (No.9), 45-48.
8. Nickell, L. G. "Plant Growth Regulators"; Springer-Verlag: New York, 1982, 147 pages.
9. Putnam, A. R. Chem. & Eng. News, 1983, 61 (No. 14), 34-45.
10. Anon. Oil Crops Outlook and Situation Report. USDA-ERS Bulletin OCS-2, 1983, pp. 8-9.
11. Ory, R. L.; Conkerton, E. J.; Sekul, A. A. in "Tropical Foods: Chemistry and Nutrition", G. E. Inglett and G. Charalambous, Eds.; Academic Press: New York, 1979, Vol. 2, 447-462.
12. St. Angelo, A. J.; Kuck, J. C.; Ory, R. L. in "Enzymes in Food and Beverage Processing," R. L. Ory and A. J. St. Angelo, Eds.; ACS Symposium Series No. 47, ACS Washington, D.C., 1977, 229-243.
13. St. Angelo, A. J.; Kuck, J. C.; Ory, R. L. J. Agric. Food Chem. 1979, 27, 229-234.
14. Bahat, A.; Rimon, D.; Hartzook, A. Hassadeh 1979, 59 (No. 9), 1782-1786.
15. Hallock, D. L.; Allison, A. H. Proc., Amer. Peanut Res. Educ. Soc. 1981, 13, 53-59.
16. Inanaga, S; Onishi, Y.; Nomura, K.; Nishihara, T. Bull. Fac. Agric. Kagoshima Univ. 1980, 30, 163-172.
17. Csinos, A. S. Southeastern Peanut Farmer 1980, 18 (No. 9), 8.

18. Csinos, A. S. Southeastern Peanut Farmer 1982, 21 (No. 5), 12.
19. Velu, V.; Ramanathan, G. Oils and Oilseed J. 1981, 34 (No. 1), 23-26.
20. Rittig, F. R.; St. John, J. B.; Becker, R.; Bleiholder, H.; Feichtmayr, F.; Haedicke, E. Pestic. Sci. 1983, 14, 299-302.
21. St. John, J. B. Plant Physiol. 1976, 57, 38-40.
22. St. John, J. B. in "Biochemical Responses Induced by Herbicides," D. E. Moreland, J. B. St. John, and F. D. Hess, Eds.; ACS Symposium Series No. 181, ACS: Washington, D.C., 1982, 97-109.
23. Ashworth, E. N.; Christiansen, M. N.; St. John, J. B.; Patterson, G. W. Plant Physiol. 1981, 67, 711-715.
24. Lem, N. W.; William, J. P. Plant Physiol. 1981, 68, 944-949.
25. Jusaitis, M.; Paleg, L.; Aspinall, D. Plant Physiol. 1982, 70, 1486-1494.
26. Willemot, C.; Slack, C. R.; Browse, J.; Roughan, P. G. Plant Physiol. 1982, 70, 78-81.
27. Terlizzi, D. E.; St. John, J. B.; Christiansen, M. N. J. Amer. Oil Chem. Soc. 1982, 59, 304A, Abst. No. 274.
28. Conkerton, E. J.; Parrish, F. W.; Chapital, D. C.; Ory, R. L. J. Food Sci. 1983, 48, 1269-1271.
29. Hymowitz, T.; Collins, F. I.; Panczner, J.; Walker, W. M. Crop Sci. 1972, 12, 710-711.
30. Ory, R. L.; Conkerton, E. J.; St. Angelo, A. J.; Vinnett, C. H.; Rittig, F. R.; Schroeder, M. Proc., Amer. Peanut Res. Educ. Soc. 1981, 13, 86.

RECEIVED April 4, 1984

9

Use of Bioregulators to Control Vegetative Growth of Fruit Trees and Improve Fruiting Efficiency

MAX W. WILLIAMS

Tree Fruit Research Laboratory, Agricultural Research Service, U.S. Department of Agriculture, Wenatchee, WA 98801

The growth retardant paclobutrazol was applied to the soil at the base of 'Delicious' and 'Golden Delicious' apple trees. One application per tree in 1979 controlled terminal shoot growth for several seasons. There was little or no effect of the chemical on fruit size. Yield increased on treated trees and the fruit colored earlier because of the increased sunlight on the fruiting spurs. The high rates of paclobutrazol caused no phytotoxicity but had an adverse effect on 'Delicious' fruit shape and stem length. A complete reversal of the adverse effects was obtained by applying Promalin at 25 ppm. Promalin contains equal amounts of GA_{4+7} and 6-BenzylAdenine. By reducing excessive terminal growth, tree efficiency is increased.

Many plants produce more leaves than are needed for maximum photosynthesis, and the shade from one or two leaves markedly reduces photosynthesis in the shaded leaves (1). The control of excessive shoot growth and shading in fruit trees is a major concern of fruit growers. Many forms of dormant and summer pruning are used to allow more sunlight into the trees to increase fruitfulness and improve fruit quality. Dwarfing rootstocks are also used to improve the manageability of fruit trees, but the proper choice of rootstock/scion combination for each soil is difficult. Optimum levels of fertilizer are important to give a proper balance between vegetative growth, fruit load and return bloom. Cultural practices such as limb spreading or tying are useful to retard vigorous upright growth and increase spur development but are expensive and generally limited to training of young trees. As trees get older heavy dense canopies shade lower portions of the tree and result in low vigor spurs with inferior flowers and fruit, and growers are at a loss to know how to manage the trees. These problems have plagued fruit growers since the beginning of domestic fruit culture, and plant breeders have yet to solve the problem.

The first attempt to control the vegetative growth of fruit trees with chemicals was reported in 1963 by Batjer, Williams and Martin (2). The chemical used was referred to as B-9, which later was given the common name daminozide, and the trade name: Alar. Some excellent uses have been developed for daminozide, such as increasing spur development, flowering, fruit set, red color and firmness of the fruit (3,4). These benefits are all obtained with moderate rates of daminozide. However, very high rates of daminozide are necessary to achieve satisfactory control of shoot growth and these high rates result in excessive reduction of fruit size.

A new compound shown in Figure 1 [1,(4-chlorophenyl)-4, 4-dimethyl-2(1,2,3-triazol-1-yl)pentan-3-01] (ICI PP333) which has the common name paclobutrazol shows considerable promise for controlling excess shoot growth without seriously reducing fruit size or quality (5-8).

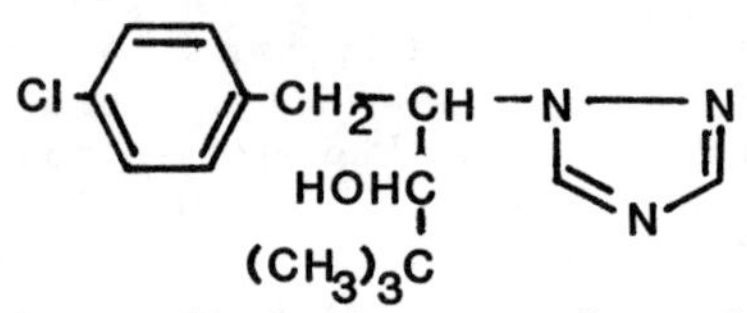

Figure 1. A new chemical compound, paclobutrazol.

Methods, Results and Discussion

In the spring (May) of 1979 trees of 'Spur Delicious' on seedling roots and trees of 'Golden Delicious' on M7 roots were sprayed with 1000 or 2000 ppm of paclobutrazol. Only about 10-20% reduction in growth occurred the year of spraying. The following season 80-90% control of shoot growth was achieved. The trees were sprayed to run off with high pressure handguns and a considerable amount of spray dripped from the foliage onto the soil under the trees. The spray material used per tree contained an equivalent of 0.5 g AI of paclobutrazol per M^2 for the 1000 ppm spray and 1.0 g AI per M^2 for the 2000 ppm spray. The major increase in response obtained the second year after spraying was related to the amount of chemical dripped onto the ground under the trees.

In the fall of 1979, 0.25, 0.5, 1.0 and 2.0 g AI per M^2 rates of paclobutrazol were applied directly to a 10 sq M area of sandy loam soil around the base of vigorous 25-year-old trees of 'Spur Delicious' on seedling roots, and 'Golden Delicious' on M7 roots. Three trees of each cultivar were treated with each rate of paclobutrazol. In 1980 only minor growth control occurred, in 1981 the full effect of treatment was manifested. The amount of growth control ranged from 10 to 90% (Table I). Treatments at the higher rates increased fruit set and final yield without reducing fruit size (Table I). Part of the increase in yield on treated trees was due to an increase in resistance to frost imparted by paclobutrazol. Spring frosts occurred in the test block in 1981 and in 1982. No hand or chemical thinning was done in 1981, thus the smaller crop loads that resulted in 1982 (Table I). Some low vigor limbs on the treated 'Spur Delicious' trees bore flattened fruit with short stems. The latter was especially true at the higher rates of paclobutrazol. Fruit quality determinations such as firmness, acidity and soluble solids were made on samples from treated and control trees. The treatments had no significant affect on the quality parameters measured (Table II). Quality tests on fruit treated in 1982 verified the 1981 data (not shown). Without any renewal treatment, the control of shoot growth in 1983 was nearly 100% from the 2 g per M^2 concentration applied in the fall of 1979.

Table I. Effect of Ground Application of Paclobutrazol Made in Fall of 1979 on Growth Measured in 1981

Treatment[a]	Fruit weight (g)	Fruit firmness (lbs)	Terminal shoot growth (% of growth)	Yield 1981 (35# boxes)	Yield 1982 (35# boxes)
		Red Delicious			
Check	166	15.51	100	7.1	4.9
0.25 g/M^2	185	15.44	95	7.5	5.6
0.50 g/M^2	174	15.16	80	10.4	7.9
1.00 g/M^2	186	15.08	40	12.5	8.3
2.00 g/M^2	179	15.42	10	15.4	6.9
		Golden Delicious			
Check	204	10.86	100	14.6	4.6
0.25 g/M^2	202	10.97	95	7.1	6.5
0.50 g/M^2	210	10.88	85	14.6	6.5
1.00 g/M^2	209	10.07	40	18.8	5.6
2.00 g/M^2	220	11.26	10	20.8	3.5

[a]Measurements made on 20 fruit and 10 shoots per tree.

In a separate trial in 1980, three vigorous 25-year-old trees of 'Top Red Delicious' on seedling roots planted 20' x 20' were treated by applying paclobutrazol at a rate of 1.0 g per M^2 in a band 2-feet wide just inside the outer edge of the tree canopy. The total area treated per tree equaled 15 M^2. Again the first year after treatment little or no reduction in growth occurred. However, in 1982, 90% control of shoot growth was achieved. The fruit were harvested from the treated and adjacent control trees and graded on a commercial fruit packing line. There were 10% more Extra Fancy fruit, because of the increased red color, and 41% more fruit of size 100 and larger from the treated trees. The percentages of cull fruit were 0.06 and 0.14, respectively, for the treated and control fruit.

Trees of 'Granny Smith' on M26 roots in their 6th leaf were treated using ground applications of paclobutrazol. The 0.25 g and 0.5 g per M^2 rate applied to an area of 1 M^2 around the trunk gave adequate growth control without seriously affecting fruit size (data not shown). The 1.0 and 2.0 g/M^2 rates on all cultivars

Table II. Effect of Paclobutrazol Fall Ground Application in 1979 on Growth and Quality Measurements of Apples in 1981

Treatment[a]	Vegetative growth (% of check)	Fruit weight (g)	Fruit size circ. (cm)	Soluble solids (%)	pH	Malic acid (%)	Fruit firmness (lbs)	Shape L/D ratio
			Red Delicious					
Checks	100	166	22.47	10.4	3.98	0.182	15.51	1.03
0.25 g/M^2	95	185	22.96	11.0	3.93	0.193	15.44	1.07
0.50 g/M^2	80	174	22.66	11.4	3.82	0.182	15.16	1.06
1.00 g/M^2	40	186	23.35	11.0	3.95	0.179	15.08	1.04
2.00 g/M^2	10	179	23.06	11.4	3.95	0.179	15.42	1.02
			Golden Delicious					
Checks	100	204	24.71	12.0	3.68	0.368	10.86	0.98
0.25 g/M^2	95	202	24.58	11.6	3.48	0.372	10.97	1.00
0.50 g/M^2	85	210	25.18	12.0	3.66	0.354	10.88	0.97
1.00 g/M^2	40	209	25.25	11.8	3.62	0.328	10.07	0.91
2.00 g/M^2	10	220	25.61	12.0	3.50	0.317	11.26	0.97

[a] Measurements made on 20 fruit or 10 shoots per tree.

tested were excessive treatments, especially when applied to trees on dwarfing rootstocks or to trees in a low state of vigor.

The flat fruit and short stem effects of paclobutrazol on apple were reversed with applications of cytokinin and gibberellins at bloom time (9). The cytokinin application on apples also increased leaf size (Table III). The results were satisfactory when only a moderate amount of reversal was needed. With 'Spur Red Delicious' 50-70% control of shoot growth may be the ideal amount of vegetative growth control to achieve to avoid any adverse fruit shape problem. With standard nonspur trees like 'Top Red Delicious', 'Golden Delicious' and 'Granny Smith', it is possible to achieve adequate control of shoot growth without seriously affecting fruit shape. Table IV indicates the effect of paclobutrazol treatments followed by the application of Promalin on fruit shape expressed as length to diameter ratios.

Table III. Average Area and Weight of Spur Leaves from 'Delicious' and 'Granny Smith' Apple Trees Treated With paclobutrazol and Promalin, 1983

Treatment	Spur Delicious Leaves Area (cm^2)	Spur Delicious Leaves Weight (g)	Granny Smith Leaves Area (cm^2)	Granny Smith Leaves Weight (g)
Control	23[a]	0.5	44	0.8
PP333 20g AI	24	0.7	34	0.8
PP333 + Promalin	33	0.9	56	1.2

[a]25 largest spur leaves per tree, 3 trees per treatment.

Table IV. Fruit L/D Ratios of 'Delicious', 'Golden Delicious', and 'Granny Smith' Apple Trees Treated with Paclobutrazol and Promalin, 1983

Treatment	Delicious	Golden Delicious	Granny Smith
Control	0.95	0.93	0.92
PP333 20g AI	0.89	0.96	0.88
PP333 + Promalin	1.16	1.06	1.03

'D'Anjou' pear trees with low, medium and high nitrogen (N) levels were treated with 2.0 g/M^2 of paclobutrazol. Six trees per treatment were observed for number of fruit set, insect damage, and overall fruit quality. The number of fruit per tree for the control, low N + PP333, medium N + PP333 and high N + PP333 treatments was 275, 414, 409, 497, respectively. Fruit finish was improved with treatment and less mite and psylla damage occurred. The results on insect damage and physiological disorders are reported elsewhere by Raese and Burts (10).

Fruit size was slightly smaller on treated pear trees because of the significant increase in fruit load. Fruit shape and stem length were affected by the high rates of paclobutrazol. The stems were short and the fruit were more compact. With shorter stems there was less stem puncture damage. At the more moderate application rates fruit shape and stem length were less affected.

Because of the increase in fruit set and the reduced amount of foliage on 'Anjou' pears, a 25 ppm gibberellic acid (GA_3) spray was applied to limb units on different trees at petal fall and at 4 weeks after full bloom in an attempt to increase leaf and fruit size. The petal fall spray increased leaf size by about 20% but did not affect fruit size. The later GA_3 spray at 4 weeks after bloom increased leaf size by about 10% and fruit size was dramatically increased. Fruit shape was not affected by the GA_3 spray applied at 4 weeks after bloom. The use of paclobutrazol to reduce excess shoot growth followed by GA_3 to increase fruit size greatly increased the fruiting efficiency of 'Anjou' pear trees (Table V).

Table V. Effect of 1980 Fall Treatment with Paclobutrazol at 20 g AI/100 Square Feet on 'Anjou' Pear Fruit Size and Volume Measured in 1982

Treatment[a]	Fruit circumference	Fruit volume	Volume of fruit/cm^2 limb x-section
Control	17.3	87.9	104
PP333	16.6	77.3	156
PP333 + 25 ppm GA_3	20.0	135.1	294

[a] GA_3 applied 4 weeks after full bloom, 1982.

Conclusions

The ability to control excessive shoot growth on apple and pear trees will be of considerable benefit to the fruit industry. Some of the benefits are: (1) less pruning required; (2) increased fruit set; (3) easier control of insects and diseases; (4) stronger spurs in lower half of trees because of increased exposure to light; (5) higher pack out of Extra Fancy red cultivars of apples because of more exposure to light; (6) easier harvesting with fewer limb rubs and handling bruises; and (7) higher pack out on 'Anjou' pears because of fewer stem punctures, limb rubs and handling bruises.

The mode of action of paclobutrazol is in the inhibition of the synthesis of gibberellins (GA) (11). GA promotes shoot growth in most plant species. By inhibiting the natural production of GA and adding GA back to the plant it is possible to have complete control of vegetative growth and fruiting. With some fruit species such as pear, cherry and peach, it is possible to increase the volume of fruit by more than 100% by treating with paclobutrazol to control excessive vegetation growth, followed by an application of GA or GA plus cytokinins to increase leaf and fruit size. These techniques have great potential for increasing the productivity of many horticultural crops.

Literature Cited

1. Heinicke, Don. R. Proc. Amer. Soc. Hort. Sci. 1966, 88, 1-8.
2. Batjer, L. P.; Williams, M. W.; Martin G. C. Proc. Wash. State Hort. Assoc. 1963, 59, 105.
3. Williams, M. W.; Batjer, L. P.; Martin, G. Proc. Amer. Soc. Hort. Sci. 1964, 85, 17-9.
4. Bartram, R. D.; Olsen, K. L.; Williams M. W. HortScience 1971, 6, 240-1.
5. Williams, M. W. Proc. PGR Society of America. 1982.
6. Quinlan, J. D.; Webster, A. D. XXIst International Horticulture Congress, Hamburg, Germany, 1982, 2, 1071.
7. Quinlan, J. D.; Webster, A. D. XXIst International Horticulture Congress, Hamburg, Germany, 1982, 1, 1285.
8. Williams, M. W.; Edgerton, L. J. Acta Horticulturae 137, Growth Regulators, 1983.
9. Curry, E. A.; Williams, M. W. HortScience 1983, 18, 214-5.
10. Raese, J. Thomas; Burts, E. C. HortScience 1982, 18, 212-4.
11. Hedden, P., unpublished.

RECEIVED February 22, 1984

Sucrose Increases with Bioregulators

LOUIS G. NICKELL

Velsicol Chemical Corporation, Chicago, IL 60611

The use of bioregulators for sucrose enhancement in sugarcane is now a commercially established cultural practice. The effectiveness of commercial products as well as highly active product candidates is a function of numerous variables including: climate, variety, crop chronological age, crop physiological age, phytotoxicity, and fertilizer status. The number and diverse chemical nature of the chemicals already found to be active as sucrose enhancers in sugarcane suggests that there are several modes of action for these bioregulators.

Sugarcane (Saccharum officinarum L.) has been known from ancient times, long before the Christian era. Its original home, for many years in dispute, is now believed to be New Guinea. The dispersal of cultivated forms of sugarcane is closely related to the ancient migrations which covered a large part of the world. Sugarcane was one of the first tropical crops to be adapted to large-scale farming. The length of the growing season varies from less than 9-10 months in Louisiana (frost dates setting the time limits), to two years in Peru and South Africa, and two years or more in Hawaii. Most sugarcane is grown in 14- to 18-month plant crops followed by 12-month ratoon crops. Cane sugar is produced commercially in over 70 countries, territories, and island groups, generally within a band around the world bounded by 35°N and 35°S latitudes (1).

Plant growth regulators have been used in the sugarcane industry worldwide for over two decades to increase the recoverable yield of sucrose in sugarcane (2-7). The first commercial success was in the prevention of flowering (8), followed by the application of gibberellic acid for the increase of stalk elongation which ultimately resulted in increased sugar production (9, 10). Currently, the primary interest centers around the use of chemicals for the control of maturation--the so called chemical "ripeners." All of these uses are now on a commercial basis as is the on/off use of paraquat as a desiccant. Research has been successful also in affecting both germination of the vegetative "seed pieces" used for propagation and

0097-6156/84/0257-0101$06.00/0

the tillering of young plants of sugarcane. Because of the importance of and the success with ripening control, these latter two uses have not been pursued rigorously. Consequently, they are not yet at the commercial stage. This brief introduction indicates that the sugar industry, in many places throughout the world, and especially in Hawaii, is using chemicals of the PGR type at almost every stage of development of the crop (3-6).

Ripening

Compounds that affect crop metabolism, particularly those that regulate crop maturity, are especially likely to have a dramatic impact on agriculture in the years ahead. Several such compounds already are used commercially and their success is in large part responsible for increased interest in plant growth regulators in agriculture. Most of the compounds used on economic crops have a direct or indirect effect on final yield, on quality, or on both.

Ripening is considered one of the most important aspects of sugarcane production--from both a research and an operational point of view. To say that the phenomenon of cane ripening is extremely complex would be, at best, a gross understatement. Studies on this use of plant growth regulators have appeared in the literature spasmodically since 1949. The first material reported to be effective was 2,4-D (11). This was followed by studies with maleic hydrazide, triiodobenzoic acid, dalapon, CMU, DCMU, EDTA, Trysben, Pesco 1850 (a mixture of MCPA and Trysben), as well as a number of enzyme inhibitors and metabolic inhibitors. No large-scale program was launched, however, until basic studies on the effects of defoliation on translocation in sugarcane had furnished a solid basis for such a program (12, 13). The screening test used is a very simple one, consisting of adding the test material by pipette or by needle and syringe into the whorl of leaves at the top of the sugarcane stalk, which is field-grown and almost at the stage of normal maturation. At a specified time or times (4, 5, and/or 8 weeks) after application of the test material, 5-20 stalks are harvested, analyzed and compared with an untreated group of stalks. The effectiveness of a test compound as a ripener is based on its ability to increase the quality of the treated stalks in two major parameters for sugar production (juice purity and sugar as a percent of field cane weight)(14).

Although attempts were made for several decades to control the ripening of sugarcane by the use of chemicals, no concerted effort was made until the start of a research program in Hawaii in the early 1960's (3, 14). This effort was soon joined by investigators in Australia (15), and Trinidad (16). The initial success resulted in extensive field testing throughout the sugarcane world (17), especially in Mauritius, South Africa, India, Brazil, the Philippines, Taiwan, Guyana, Colombia, Puerto Rico, Australia, Jamaica, Trinidad and the mainland U.S.A. Originally, very few companies were involved; chemicals used were primarily those available from chemical supply houses together with the few materials synthesized by research organizations in the sugarcane industry around the globe. The initial success led a number of companies to become interested in supplying chemicals for evaluation. This total effort has resulted in a surprising number of chemicals that increase the sucrose content of sugarcane at harvest. Most of those compounds that have met with sufficient success to have information published about them are given in Table I. Many compounds were never developed beyond the initial screening stages.

The first material seriously considered as a candidate ripener for increasing sucrose yield of sugarcane was the dimethylamine salt of 2,3,6-trichlorobenzoic acid. Because of a number of technical, environmental, and legal problems, this material did not prove successful commercially (18). Nevertheless, it served as a standard for comparison in screening tests aimed at finding better sugarcane ripeners. It continued to be the standard for comparison until the registration of the first ripener for sugarcane in the United States. This compound is N,N-bis(phosphonomethyl)glycine, known generically as glyphosine and marketed as the product Polaris (19, 20).

Glyphosine has been evaluated over a period of several years in Hawaii and other sugar-producing areas (19), and has given substantial gains--about 5 to 15 percent increase in yields--which is an increase of 1 ton per acre or more when applied to certain varieties grown on the rainy coasts of the Island of Hawaii. More recent work has shown that varieties previously thought to be nonresponsive to this ripener have been found to respond positively when surfactants are added to the formulation (2). Similarly, it has been found to be effective on irrigated lands when surfactants are added. Glyphosine treatment results in a reduced rate of terminal cane growth, but how this relates to its mode of action has not yet been established.

Until late 1980, glyphosine was the only compound registered as a sugarcane ripener in the United States. In the fall of 1980, phosphonomethyl glycine was registered as a ripener for sugarcane. This compound, known generically as glyphosate (21, 22), is marketed as the product Polado; it is the sodium salt of the same compound that is the active ingredient of the herbicide Round-Up. Glyphosate is almost an order of magnitude more active than glyphosine. Glyphosate formulations improve the sucrose content over a wide range of climatic conditions, are less cultivar-specific, and the ripening response they induce in sugarcane is more consistent and rapid than that obtained with glyphosine.

Three other chemicals: Ripenthol, chlormequat, and mefluidide have been registered under experimental labels in the United States for field evaluation as commercial ripeners.

Ripenthol, the monoamine salt of Endothall, was one of the first materials found to have significant activity on sugarcane in Hawaii (14). Numerous relatives of this compound were tested in the early screening stages, and it was found that, while the acid itself had very low activity, amine salts were more active than disubstituted amines. Ripenthol (also known as Hydrothol) has considerable phytotoxic activity and, because of this, care must be taken in its application, especially to avoid drift when applied by air.

Chlormequat (2-chloroethyl-trimethylammonium chloride), also known as Cycocel, is among the most widely used plant growth regulators in the world on crops other than cane. It has been evaluated on more than 1,000 acres of sugarcane in Hawaii, but preliminary results suggest that its activity might be too low to be commercially successful.

TABLE I. Sugarcane Ripening Compounds*

Compound	Common Name	Trade Name/ Code Designation
p-aminobenzenesulfonyl urea	-	-
2-amino-6-methyl benzoic acid	6-amino-o-toluic acid	ACR-1308
aminoethylphosphonic acid	-	AMPA
6-aminopenicillanic acid	-	6-APA
4-amino-3,5,6-trichloropicolinic acid	picloram	Tordon
ammonium ethyl carbamoyl phosphonate	-	-
ammonium isobutyrate	ammonium isobutyrate	AIB
N-benzoyl-N-(3,4-dichlorophenyl)aminopropionic acid	-	-
N,N-bis(phosphonomethyl)glycine (39, 41)	glyphosine	Polaris
bis(N,O-trifluoroacetyl)-N-phosphonomethyl glycine	-	-
2-bromobenzylphosphonic acid	-	-
5-bromo-3-sec-butyl-6-methyl-uracil	bromacil	Hyvar X
butyl-2-[4-(5-trifluoromethyl-2-pyridinyloxy)phenoxy] propanoate (42)	fluazifop-butyl	Fusilade
2-chlorobenzoic acid	-	-
2-(chloro-2,4-dimethoxyphenyl)-thioureido-3-cyano-4,5-tetramethylenethiopene	-	-
2-chloroethylaminodi(methylphosphonic acid)	-	-
2-chloroethylphosphonic acid (40)	ethephon	Ethrel, Cepha
2-chloroethyltrimethylammonium chloride (40)	chlormequat	Cycocel, Lihocin
5-chloro-2-thenyl-tri-n-butyl-phosphonium chloride	-	CHE-8728
3-cyclohexene-1-carboxylic acid	tetrahydrobenzoic acid	-
2,6-dichlorobenzylphosphonic acid	-	-
2,3-dichloro-6-methylbenzoic acid	-	-
2,4-dichlorophenoxyacetic acid (40)	2,4-D	Brush Killer, Weedone, Super-D, Weed-B-Gon, Weedtrol
2,2-dichloropropionic acid	dalapon	Dowpon, Basfapon

N-(2,3-dihydroxy-1-propyl)-N-phosphonomethyl glycine, disodium salt	-	-
Diisobutylphenoxyethoxy-ethyl-dimethylbenzyl ammonium chloride	-	-
2-(β-dimethylamino-ethoxy)-4-(3',4'-dichlorophenyl)-thiazole hydrochloride	-	-
dimethylarsenic acid	cacodylic acid	Phytar 138
N,N-dimethylglycine	-	-
3-[2-(3,5-dimethyl-2-oxocyclohexyl]-2-hydroxy-ethyl)glutarimide	cycloheximide	Actidione
N-[2,4-dimethyl-5-[[(trifluoromethyl)sulfonyl] amino]phenyl]acetamide	mefluidide	Embark
diphenylchlorophosphate	-	-
2-(1-ethoxyimino)-butyl)-5-(2-(ethylthio)-propyl)-3-hydroxy-2-cyclohexene-1-one (43)	-	-
ethyl-N-(2-cyanomethyl)-N-ethoxy phosphonomethyl-glycinate, monosodium salt	-	-
2-formyl-4-chlorophenoxyacetic acid	-	SAE1-517
hexadecyltrimethylammonium bromide	cetyltrimethylammonium bromide	Cetrimide
furfuryl 2-methoxy-3,6-dichlorobenzoate	-	-
ethyl 2-[4-(5-bromo-2-pyridyloxy)phenoxy]propionate	-	-
1-hydroxy-1,1-ethane diphosphonic acid	-	-
4-hydroxy-3-methoxy benzaldehyde	vanillin	-
imidodicarbonic diamide	carbamyl urea	Biuret
isochlortetracycline	isoaureomycin	-
laurylmercaptotetrahydropyrimidine (40)	-	-
β-mercaptovaline	penicillamine	Cuprimine
N-[(4-methoxy-6-methyl-1,3,5-triazin-2-yl)-amino carbonyl]-2-chlorobenzene sulfonamide	-	-
N-[(4-methoxy-6-methylamino-1,3,5-triazin-2-yl)-amino-carbonyl]benzene sulfonamide	-	-

Continued on next page

Table I. Continued

methyl-3,6-dichloro-o-anisate (40)	disugran	Racuza
methyl-2-[4-(2,4-dichloro-phenoxy)phenoxy]propionate	-	-
methyl-2[(4,6-dimethylpyrimidin-2-yl)amino carbonyl] aminosulfonyl benzoate (44)	-	-
7-methyl indole	-	PP-757
3-(2-methylphenoxy)pyridazine	-	Credazine
2-methyl-1-propanol	isobutanol	-
methylsulfanil-yl-carbamate	asulam	Asulox
methyl-2-(ureidooxy)propionate	-	DA-5
2-(p-methoxybenzyl)3,4-pyrolidine-diol-3-acetate	anisomycin	Flagecidin
7-oxabicyclo-2(2,2,1)-heptane-2,3-dicarboxylic acid, monoalkylamine salt	endothall	Ripenthol
n-pentanoic acid	n-valeric acid	-
6-phenoxyacetamido-penicillanic acid	penicillin V	Pen Vee
N-(2-phenoxyethyl)-N-propyl-1H-imidazole-1-carboxamide	-	BTS 34-273
N-phenylphosphinylmethyliminodiacetic acid-N-oxide	-	-
N-phenylsulfonamido-N-phosphonomethyl glycine	-	-
phosphonic acid, (2,2,2-trichloro-1-hydroxy-ethyl)-bis-2-(2-hydroxypropoxy)-1-methylethyl ester	-	-
N-phosphonomethylglycine	glyphosate	Polado
poly[oxyethylene(dimethylimino)ethylene (dimethylimino)ethylene dichloride]	-	-
sucrose ester of 2-methoxy-3,6-dichlorobenzoic acid (45)	-	-
tetrahydrofuroic acid hydrazide	-	-
1,2,4-triazine-3,5(2H,4H)-dione	6-azauracil	-
N-trichloracetylaminomethylene phosphonic acid	-	-
2,3,6-trichlorobenzoic acid, dimethylamine salt	2,3,6-TBA	Trysben
3-(trifluoromethyl-sulfonami do)-p-aceto-toluidide	fluoridamid	Sustar
--	bacitracin	Bacitracin
--	mineral oil	-
--	-	Tergitol NPX
--	-	Tween-20

--	-	LFA 2129
--	-	ACR 1093
--	-	Streptocyclene (40)

*Table extracted from "Plant Growth Regulators: Agricultural Uses", Nickell, L. G., 1982; and Nickell, L. G. In "Plant Growth Regulating Chemicals", Nickell, L. G., Ed., CRC Press, 1983, Vol. I, Chapter 7.

Mefluidide, also known as Embark, is being tested at the present time under an experimental label in Hawaii, the Philippines, and certain other countries.

The ethylene-producing compound ethephon is used commercially on sugarcane in South Africa (23, 24) and Rhodesia (25). The effectiveness of ethephon as a sugarcane ripener has not been comparable to that of glyphosine in some areas of the world, although its growth effects generate an increase in yield. These effects are being evaluated in research programs in the sugar industry.

It has been known since the early days of ripening evaluations that ethephon can be active as a ripener. The fact that the results obtained over a period of time have been variable between tests (particularly under different conditions such as season) and that the activity usually was not as great as that of other available materials throughout the testing time, suggests either that ethephon is not a competitive ripener or that it has other activity in addition to ripening which complicates the picture. The first suggestion of what this complication might be was presented by Jaramillo and co-workers (26), who showed increased cane weights in tests initiated to evaluate ethephon as a ripener. This work was conducted in Colombia and reported in 1977. Ethephon was included in tests over several years in Hawaii comparing it with other potential ripeners. The results showed, under the conditions of the tests, that the quality of the cane following ethephon treatment was depressed, which is an indication of growth rather than ripening. The harvest results confirmed this showing consistently increased stalk fresh weight between 8 and 10 weeks following treatment (27). This led to serious consideration of ethephon as an enhancer of cane weight to be followed by a different ripener as a sequential treatment. The results of these tests over a two-year period using glyphosate following ethephon treatment showed consistent increases in cane growth attributable to ethephon. Although the reasons for this ethephon-induced growth stimulation have not been determined, it has been likened to that postulated for gibberellic acid. In addition to increasing the distance between photosynthetically active leaves, the leaf length is reduced as a result of ethephon treatment. These investigators suspect that the shorter, more erect, more widely spaced leaves may allow for greater light penetration into the canopy and account for the increased level of dry matter production (27-29).

Studies confirming the increased growth effects from ethephon application and comparing the results to those obtained by similar treatment with gibberellic acid have been reported also from South Africa (30). The similarity to results obtained with gramineae (31) is pointed out.

The number, and very chemical nature, of compounds that are active as sugarcane ripeners suggests that there are several modes of action to enhance the ripening of sugarcane and that the active compounds might fall into any of these several classifications. There are also varietal differences, as well as differences due to (a) the fertilizer status (particularly nitrogen), (b) the age of the crop and its condition, (c) the climate (both during the growth of the crop and prior to harvest), (d) the physiological state of the cane, and (e) the purity of the juice in the young growing tops. These variables, and probably many others, suggest that there is room for a number of ripeners on sugarcane. Additional variables to be considered are (a) phytotoxicity of the ripeners, (b) the cost effectiveness of the compound under consideration, and (c) the effects on the processing of sugarcane.

The effects of ripeners on the growth of subsequent ratoon crops is an important factor because of potential adverse effects on tillering and regrowth of the ratoon crop. If this occurs, it also raises the question of residues in the stubble.

Studies in Florida showed both glyphosine and glyphosate to reduce the regrowth of two test cultivars in commercial scale tests, with glyphosate causing a greater reduction than glyphosine (32). Ratoon regrowth studies in Hawaii showed poorer growth at higher glyphosate rates (1 and 2 lb per acre), but no adverse effect at the recommended rate (0.5 lb per acre)(21). Results in Louisiana were similar to those in Hawaii (33).

Early work with glyphosine in Louisiana produced numerous non-responsive fields (34). Studies to determine the agent(s) or factor(s) which might reduce the level of response showed that cane infected with ratoon stunt disease did not respond to glyphosine treatment (35).

Recent comparative studies with four active sugarcane ripeners: glyphosine, glyphosate, ethephon, and mefluidide, have demonstrated two models for increasing sucrose per stalk in sugarcane. Glyphosine and glyphosate increase sucrose per stalk by increasing the partitioning of dry matter toward sucrose storage and away from fiber production. Both mefluidide and, to a much larger extent, ethephon increased the production of sucrose per stalk by increasing the total amount of dry matter produced, with a greater portion incorporated as fiber and a lesser amount as sucrose (28, 29, 36). The possibility of two ripeners, acting through different modes of action, having additive effects is quite reasonable, although little has been done to substantiate this theory.

Probably because glyphosine was the first registered sugarcane ripener, the use of this compound had a meteoric rise, at least in the cane areas of the United States. From its registration in 1972 and experimental use on a few hundred acres, it reached over 60,000 acres in Hawaii by 1977. A similar situation occurred in Florida, increasing from 178 acres in 1972 to more than 46,000 acres in 1979 (37). Because of the greater activity and considerably lower costs with glyphosate, it is expected that glyphosine will be replaced quickly. Recommended dosage for glyphosine is about 4 lbs active per acre whereas for glyphosate it is about 0.5 lb active per acre.

Rostron (24) found ethephon to be much more effective than glyphosine in southern Africa, whereas the reverse was found to be true in Hawaii and other places where the two have been compared. Glyphosate has been found to be quite active in the sugarcane growing regions of southern Africa, and it is expected that it will be an effective competitor for ethephon.

Interest seems to be declining for three materials that received early consideration: chlormequat, disugran, and Ripenthol. In several major sugar growing areas, mefluidide appears to be losing popularity, but it is being more seriously evaluated in tropical countries such as the Philippines.

The financial return to the grower is substantial through the use of ripeners; increased sugar yield produced by such compounds can be as much as 15-20 percent, depending on the variety of sugarcane treated as well as on prevailing weather and soil conditions. Chemical control of maturation in sugarcane is now an established practice. In fact, it is so established that many research organizations in the sugar industry are shifting part of their efforts to looking at other stages in the development of the sugarcane crop for additional potential uses and times of chemical treatment. It is thought ethephon might fall into this sort of economic use, not unlike that found for gibberellic acid.

Concluding Remarks

Traditionally, the sugar industry worldwide has supported a multi-pronged investigation of the activity of chemicals on most of the steps of cane development, from germination through ripening and harvest. While the use of plant growth regulators is still in its infancy, success to date with ripening alone in terms of yield increases greater than 10 percent substantiates the belief that the regulation of crop growth and metabolism may result in one of the most important quantitative gains yet achieved in agriculture (38). The monumental task of producing raw materials to supply the world's food and to supplement its energy requirements may depend to a large degree on achievements of this magnitude in a wide range of crops.

Hawaii, with its high costs of operation and high yields of sugarcane, nonseasonal environmental conditions, and the necessity to harvest the year-round, can afford high-priced chemicals. In other cane-producing countries, this may not be the case. For example, Australian investigators were among the first to study the use of chemical ripeners. In fact, during the early 1960's, there was a cooperative program between Hawaiian and Australian workers on this subject. This was not pursued when it was realized that Australian conditions, with cool and dry weather at harvest, were for the most part conducive to excellent natural ripening. In other cane-growing areas in the world, however, as more is learned about the relationship of a given chemical to the process which it affects, as sugarcane agronomy improves and its economy become more favorable, the use of chemical ripeners will undoubtedly become more widespread; essentially the same can be said for the use of gibberellins and chemicals for flower control and other uses.

Historically, agricultural research has been primarily concerned with improvement of total crop yield by the removal of obstacles to optimize production. Now that many of these obstacles can be overcome with herbicides, pesticides, fertilizers, irrigation, and improved management practices, the stage is set for further yield increases by the use of sophisticated techniques of physiological manipulation of the plant and its metabolism.

Literature Cited

1. Nickell, L. G. In "Ecophysiology of Tropical Crops"; Alvim, P. de T.; Kozlowski, T. T., Eds.; Academic Press: New York, 1977; p. 89-111.

2. Nickell, L. G. Bull. Plant Growth Regulators, 1974, 2(4), 51.
3. Nickell, L. G. Outlook on Agriculture, 1976, 9, 57.
4. Nickell, L. G. In "Plant Growth Regulators"; Stutte, C. A., Ed.; ACS ADVANCES IN CHEMISTRY SERIES No. 159, American Chemical Society; Washington, D.C., 1977; p. 6-22.
5. Nickell, L. G. In "Plant Growth Substances 1979"; Skoog, F., Ed.; Springer-Verlag: Berlin-Heidelberg-New York, 1980, p. 419-425.
6. Nickell, L. G. In "Chemical Manipulation of Crop Growth and Development"; McLaren, J. S., Ed.; Butterworth & Co.: London, 1982, p. 167-189.
7. Nickell, L. G. "Plant Growth Regulators: Agricultural Uses"; Springer-Verlag: Berlin-Heidelberg-New York, 1982.
8. Tanimoto, T.; Nickell, L. G. Proc. 12th Int. Soc. Sugar Cane Technol., 1965, p. 113.
9. Tanimoto, T.; Nickell, L. G. Rpts. 1966 Haw. Sugar Technol., 1967, p. 184.
10. Tanimoto, T.; Nickell, L. G. Rpts. 1967 Haw. Sugar Technol., 1968, p. 137.
11. Beauchamp, C. E. Proc. 23rd Ann. Mtg. Assn. Tecn. Azucareros, Cuba, 1949, p. 55.
12. Hartt, C. E. Rpts. 1963 Haw. Sugar Technol., 1964, p. 151.
13. Hartt, C. E.; Kortschak, H. P.; Burr, G. O. Plant Physiol., 1964, 39, 15.
14. Nickell, L. G.; Tanimoto, T. T. Rpts. 1965 Haw. Sugar Technol., 1966, p. 152.
15. Glasziou, K. T. Rpt. David North Research Centre, 1964, p. 48.
16. Vlitos, A. J.; Lawrie, I. D. Proc. 12th Int. Soc. Sugar Cane Technol., 1965, p. 429.
17. Humbert, R. P. World Farming, 1974, 16(12), 25.
18. Nickell, L. G.; Tanimoto, T. T. Rpts. 1967 Haw. Sugar Technol., 1968, p. 104.
19. Nickell, L. G.; Takahashi, D. T. Rpts. 1971 Haw. Sugar Technol., 1972, p. 73.
20. Ahlrichs, L. E.; Porter, C. A. Proc. 11th British Weed Control Conf., 1972, p. 1215.
21. Osgood, R. V. Rpts. 1976 Haw. Sugar Technol., 1977, p. 60.
22. Osgood, R. V. Proc. 7th Plant Growth Regulator Working Group, 1980, p. 149.
23. Clowes, M. S. J.; Wood, R. A. Proc. 52nd So. African Sugar Technol. Assn., 1978, p. 166.
24. Rostron, H. A. Proc. 16th Int. Soc. Sugar Cane Technol., 1977, Vol. 2, p. 1605.
25. Sweet, C. P. M. Proc. 16th Int. Soc. Sugar Cane Technol., 1977, Vol. 2, p. 1619.
26. Jaramillo, H.; Schuitemaker, F.; Garcia, C. Proc. 16th Int. Soc. Sugar Cane Technol., 1977, Vol. 2, p. 1931.
27. Osgood, R. V.; Teshima, A. Ann. Rpt. Exp. Sta. Haw. Sugar Planters' Assn., 1979, p. 40.
28. Osgood, R. V.; Teshima, A. Ann. Rpt. Exp. Sta. Haw. Sugar Planters' Assn., 1980, p. 33.
29. Osgood, R. V.; Teshima, A. Proc. 7th Plant Growth Regulator Working Group, 1980, p. 150.

30. Clowes, M. S. J. Proc. 54th So. African Sugar Technol. Assn., 1980, p. 146.
31. Van Andel, O. M.; Verkerke, D. R. J. Exp. Bot., 1978, 29, 639.
32. Rice, E. R. Sugar y Azucar, 1981, 76(6), 29.
33. Legendre, B. L.; Martin. F. A.; Dill, G. M. Proc. 7th Plant Growth Regulator Working Group, 1980, p. 148.
34. Legendre, B. L.; Martin, F. A. Proc. 6th Amer. Soc. Sugar Cane Technol., 1977, p. 62.
35. Martin, F. A.; Steib, R. J.; Dill, G. M. Proc. 17th Int. Soc. Sugar Cane Technol., 1980, Vol. 1, p. 614.
36. Osgood, R. V.; Moore, P. H.; Ginoza, H. S. Proc. 8th Plant Growth Regulator Soc. Amer., 1981, p. 97.
37. Rice, E. R.; Holder, D. G.; DeStefano, R. P. Sugar Journal, 1980, 43(5), 23.
38. Nickell, L. G. Chem. & Eng. News, 1978, 56(41), 18.
39. Holder, D. G.; DeStefano, R. P.; Stacy, J. D.; Beardsley, J. W. Proc. 8th Amer. Soc. Sugar Cane Technol., 1978, p. 56.
40. Singh, R. G.; Khan, H. A.; Lal, K.; Bhargava, A. Proc. 46th Ann. Conv. Sugar Technol. Assn. India, 1982, p. Ag 45.
41. Rao, K. C.; Thangavelu, S. Proc. 44th Ann. Conv. Sugar Technol. Assn. India, 1980, p. Ag 87.
42. Dusky, J. A. Sugar y Azucar, 1982, 77(6), 30.
43. Koerwer, J. F. U.S. Patent 4334913, 1982.
44. Levitt, G.; Smith, W. F. U.S. Patent 4362549, 1982.
45. Luteri, G. F. U.S. Patent 4345933, 1982.

RECEIVED February 22, 1984

11

Plant Growth Regulator Uses on Citrus

W. C. WILSON

Florida Department of Citrus, Lake Alfred, FL 33850

Plant Growth Regulators (PGR's) are used on citrus worldwide and active research programs are in progress in most citrus-growing countries. Principal uses for PGR's on citrus are for control of fruit maturity, reduction of fruit drop, prevention of rootstock sprouting, fruit thinning, preservation of fresh fruit peel quality, and control of the abscission process for harvesting. However, there is a tremendous potential for PGR's for freeze protection (cold hardiness), improving internal and external fruit color and quality, improving storage life of fruit, increasing kg solids/ha, increasing vitamin C, controlling vegetative growth and inducing flowering and fruit set. Because citrus is grown under a wide variety of climatic conditions, this can greatly affect the performance of a PGR chemical on a given cultivar.

A tremendous amount of research has been conducted worldwide with plant growth regulators (PGR's) on virtually every crop. Unlike herbicides (the "cousins" of PGR's), vast commercial development has not been as rapid, as the plant has been slow in yielding its growth regulating secrets. It would appear that selective killing of plants is easier than controlling their functions and activities. Nevertheless, significant breakthroughs have been achieved on many crops, including citrus, and it appears that commercial usage of these materials will increase in the future. Many reasons have been advanced as to why PGR's have not found markets as rapidly as other plant protection chemicals. My own experience using abscission agents indicates there are several reasons, including variability in chemical activity; differing climatic conditions and cultivars; basic costs of chemicals and applications; phytotoxicity and human toxicity problems.

Citrus production is limited primarily by freezing temperatures because the trees will grow under a wide variety of tropical and subtropical conditions (1). Numerous cultivars of sweet orange (*Citrus sinensis* (L.) Osbeck), grapefruit (*C. paradisi* Macf.),

0097-6156/84/0257-0113$06.00/0

mandarin (tangerine or C. reticulata Blanco), lemon (C. limon (L.) Burm. f.), lime (C. aurantifolia Swing.) and citrons (C. medica L.) are grown worldwide.

PGR spray applications reported in the numerous citations were usually chemically dilute concentrations. Spray coverage was usually thorough (complete). Concentrate sprays can reduce fixed spraying costs for many pesticides, but our experience with PGR's is that most have reduced effectiveness if concentrated (2, 3). Further, our experience with surfactants and wetting agents has shown that they are beneficial to some PGR's, but deterimental to others (2, 3 and see infra).

Principal Commercial Uses of PGR's on Citrus

Fruit Thinning. Most commercial citrus cultivars do not require thinning, particularly if the crop is destined for processing utilization. However for fresh fruit use, size, color, and other features are often important selling points. Some cultivars, mandarin in particular, tend to alternate bearing, thus producing heavy crops of small fruit one season but light crops of excessively large fruit the following season. Therefore, thinning to increase size (in the "on" year) and reduce alternate bearing is often an important horticultural consideration.

The Japanese public is very conscious of fresh fruit quality and is willing to pay a premium for it. Therefore, thinning to improve fruit quality (particularly size) of satsumas is regularly practiced (4). A leaf to fruit ratio of 25:1 is desired, and the purpose of chemical thinning is to reduce the amount of hand thinning necessary to achieve this ratio. The most effective material has been NAA (1-napthaleneacetic acid) applied 25 days following full bloom. This caused about a 30% increase in fruit drop and effectively thinned the crop (5, 6).

Thinning with NAA in Japan is affected by climatic conditions, particularly temperature (4). Best results occurred when post spray temperatures were 77°F (25°C). High humidities also increased thinning, presumably by causing increased chemical uptake.

A new growth regulator, Figaron (IZAA or ethyl-5-chloro-H-3 indazolyl-acetate), is registered for use in Japan for thinning and quality improvement of satsumas (4, 5). IZAA causes no phytoxicity. With satsuma, it is also reported to advance fruit color and to increase juice Brix (sugar content). Tests with the compound in Florida, however, have been ineffective (3).

Ethephon (2-chloroethylphosphonic acid) has also been tested for thinning satsumas in Japan (4, 6), but caused excessive defoliation. Gibberellic acid (GA) applied prior to bloom effectively thinned satsumas by decreasing flowering (4).

In the USA (Florida), Wheaton (7) reported the principal need to thin fruits has been with the cultivars 'Dancy' and 'Murcott' tangerine, both of which tend to alternately bear and produce excessively heavy crops of small-sized fruit during the "on" year. NAA was the most effective and consistent material tested on both cultivars. The auxins CPA (3-chlorophenoxy acetic acid), 2,4-D (2,4-dichlorophenoxy acetic acid) and 2,4,5-T (2,4,5-trichlorophenoxy acetic acid) were also effective thinning agents, although 2,4,5-T sometimes tended to overthin and cause phytotoxicity.

In tests with ethylene-releasing materials, ethephon was

generally effective with 'Dancy' and caused a minimum leaf drop, but 'Murcott' showed considerable sensitivity to it and over-thinning was a problem (7). GA applied prebloom provided fair thinning of 'Dancy' but was ineffective on 'Murcott'.

Fruit thinning tests in other parts of the world with mandarin-type fruits have been similar to those mentioned (8, 9, 10).

In some parts of the world, alternate bearing of the 'Valencia' orange warrants correction by fruit thinning. In Australia, thinning was accomplished with ethephon applied when fruit was 1.0-1.5 cm in diameter (about 4 weeks following bloom) (11, 12). Also, GA has proven to be an effective means to control fruiting of 'Valencia' when the spray was applied in winter prior to the "on bloom" year (12). The GA application increased the number of vegetative buds vs. flower buds, and although cropping during the "on year" was not substantially reduced in some cases, cropping during the "off year" was substantially increased. Hand-thinning studies have shown that removal of only 22% of the flowers in the "on" year would greatly reduce crop differences in the two-year cycle (13). The alternate bearing problem does not appear to be of major importance in areas where the bulk of the fruit is utilized for processing purposes (3).

Preharvest Drop Control and Storage of Fruit on the Tree. In Florida, an application of 2,4-D is recommended to prevent dropping of 'Pineapple', seedlings and 'Temple' oranges, and seedless grapefruit (14). Dilute (only) sprays should be applied in November or December. In California and other citrus-growing areas of the world, the use of 2,4-D has generally been more successful than in Florida and is reported to prevent preharvest fruit drop of most cultivars (9, 15, 16).

Because of the importance of the Japanese market to Florida, prevention of grapefruit aging and a means of lengthening the season are important considerations (17). Therefore, should inhibition of color change and prevention of peel aging of seedless (sparsely seeded) grapefruit be required, a combination of 2,4-D and GA can be used (14, 18). Similar results with export grapefruit have been reported from Australia and South Africa (19, 20, 21, 22). Although delayed or late season harvesting of 'Valencia' may depress yields the following year, this reduction has not been observed with seedless ('Marsh') grapefruit (23).

Preservation of Fresh Fruit Peel Quality. There are a number of peel disorders which can be alleviated by applications of specific plant growth regulators (24). It should be pointed out, however, that problems with peel are of primary interest to growers and shippers of fresh fruit just as internal qualities are of prime concern to processors.

With lemons grown in California, and probably most other Mediterranean climate areas of the world, peak production occurs during winter and spring when demand is low. Most lemons are harvested, stored (cured), then packed and shipped to market much later. However, during the summer when fresh lemon sales are strongest, the amount of fruit available is lowest. Therefore, it is desirable to extend the lemon harvest season by prevention of premature yellowing (senescence) of the fruit. GA applied in the fall is reported to be the most beneficial treatment to delay overall maturity (and yellowing) (25). These treatments also affect the second-year harvest pattern, probably from the influence of the GA on flowering, resulting in more trees producing fruit during summer when market demand is high (9). Benefits of the GA treatment to lemons were better flexibility in harvest patterns, a longer storage life, and a reduction in the number of small, yellow lemons (9).

Navel oranges in the Central Valley of California have rind softening problems which develop as the fruit matures and peel color changes from green to orange (9). These senescence changes can contribute to a number of other rind disorders such as rind staining, sticky rind, puffy peel, and water spot (also considered to be weather related) (24). GA treatments applied in the early fall lessen or prevent these conditions, thus allowing a more orderly fresh fruit marketing season.

Creasing affects many commercial citrus cultivars throughout the world. It is the most important single cause for rejection of 'Valencia' fruit in Israeli packinghouses, causing a 26% discard rate (26). The use of GA applied when fruit is 3-4 cm in diameter (about July, or 4 months following anthesis) caused considerable reduction in the incidence of creasing without impeding good color development. The mode of action of GA was believed to be through renewal of growth activity in the affected tissues.

In South Africa, creasing of navel orange can cause fresh fruit packinghouse losses as high as 50% near the end of the shipping season (27, 28). The recommended control measure is to apply GA when young fruit is 30-50 mm in diameter (70-100 days after anthesis).

The peel of mandarin fruit often tends to puff. In Japan, GA applied to satsuma mandarin (2 applications), reduced the amount of puffy fruit but also slowed the rate of chlorophyll degradation (29).

Fruit Size. Increased fruit size is desirable for some cultivars as small sizes often cause serious profit losses for fresh fruit packinghouses which must eliminate them in order to market fruit of a specified legal size. Mandarin-type fruits are often thinned with growth regulators for this purpose (see supra). For fruit destined for processing, however, size is usually not a consideration as increasing pounds solids per acre (kilograms solids per hectare) is paramount. An extensive review of all factors affecting fruit size and suggested methods for its improvement has recently been completed by Gilfillian (30) in South Africa.

Lemons tend to produce their heaviest crops during the winter and spring when consumer demand is low (previously mentioned). Therefore, it is advantageous to produce fruit which can be picked and stored until the summer period when demand (and prices) are high. In California (31), GA and/or additional potassium fertilizer applications are used to produce the larger, greener fruits which can be harvested and stored up to 6 months. The small, yellow, tree-ripened fruits are removed during the winter-spring period and primarily utilized for processed products.

In Australia (32), when lemons are held on the tree into the late spring, many grow to a size which is not acceptable in the marketplace. The fruit tends to be orange-yellow which is not favored by buyers. A preharvest combination application of GA and CCC applied in the fall delayed fruit coloring and arrested fruit growth, thus extending the harvest into late spring and furnishing a larger quantity of small-sized, premium-quality fruit (32).

Postharvest Treatments, Fruit Shipment and Storage. The citrus fruit is fully ripe at harvest, it contains practically no starch reserves, and it is not a good candidate for controlled atmospheric storage (33). Although it is suggested that the best place to store citrus probably is on the tree, there are occasions when it must be harvested and stored, or transported very long distances in the hold of ships. Most cultivars have a long harvest period and fruit response to plant growth regulators can vary depending on the time during the season it was harvested, the time of the day it was harvested, tree condition, rootstock, location, year-to-year weather variations, etc. In assessing the performance of plant growth regulators on citrus, some or many of these varying conditions may affect activity of the chemical.

Ethephon has been used to substitute for the effects of ethylene degreening treatments for fruit. In Florida (34, 35), dipping lemons in an ethephon solution was found to hasten the development of marketable color by 5-14 days. There was no significant effect on the amount of decay. However, degreening mandarins with ethephon usually is more successful when applied as preharvest sprays (36). In South Africa, however, postharvest ethephon dips of oranges are being used commercially in lieu of ethylene treatments (37).

The problem with lime is the opposite of lemon since it is desired to retain the green fruit color which the public associates with limes (38).

In California, as well as similar lemon-growing areas, preservation of the calyx tissue ('button') on the fruit prevents stem-end rot fungus from entering through the abscission layer tissue. This can be successfully controlled by fruit dips in aqueous solutions of 2,4-D (33), or by applications of fruit waxes containing 2,4-D or 2,4,5-T (9). Preservation of the 'button' also improves the cosmetic appearance of the fruit, and many believe this manifests its freshness to the housewife.

A very good Japanese market for grapefruit has developed in recent years. Because of the long transit time in the shiphold (about 3-5 weeks from Florida), shipping losses can, at times, be substantial. Florida grapefruit is subject to chilling injury (33), particularly fruit harvested in early fall. As previously mentioned, (14, 18, 19, 20) preharvest treatments with 2,4-D and GA can reduce peel senescence. Attempts to control chilling injury with plant growth regulators, however, have produced conflicting results (39).

The presence of seeds in citrus fruit can occasionally cause problems (40, 41). Grapefruit which are held very late into the shipping season often have sprouted seeds. Preharvest treatments with 2,4-D and GA, previously mentioned, reduced the number of sprouted seeds (17, 18), but recent research was unable to confirm this (42).

Reduction of Acidity. Improvement of grapefruit flavor has been practiced for many years in some areas through the use of arsenate (14). The most effective application period seems to be 1 to 6 weeks following bloom; however, it can be applied as late as 4 months post bloom. Arsenate causes a reduction in total acidity of grapefruit and, consequently, causes an increase in Brix/acid ratio. It is also effective on oranges, but its effect is usually so pronounced that most of the treated fruit are insipidly sweet. Most organic and inorganic arsenical compounds will cause acidity reduction in citrus fruits (3), but, so far, no nonarsenical compounds have been identified which appear to have commercial possibilities (43).

In South Africa, arsenical sprays have also been tested extensively and have been used commercially for many years (44, 45, 46). South African oranges, particularly the 'Valencia', tend to produce fruit which has high acid content; therefore, their research efforts have been aimed at reducing acidity in orange cultivars as well as grapefruit.

Recent research with arsenicals in the saline waters of the Sundays River Valley has shown that the total quantity of arsenic used can be reduced by acidifying the tank mix to pH 4 with sulfuric acid (47, 48). (Phosphoric acid (48), however, caused considerable phytotoxicity when used.) Calcium arsenate is the preferred form (46). Arsenic residues on fruit are negligible (3, 48, 49), with the largest proportion found in the peel and practically none in the pulp and juice (48, 49).

In the past, inorganic arsenic has been considered to be a carcinogen and EPA clearance (or maintenance of existing clearances) of arsenicals on food crops in the United States has been quite difficult. However, recent nutritional research throughout the world has established arsenic as an essential nutrient for certain animal species (50). Although arsenic has not been proven to be essential for humans, future research may, indeed, confirm these findings. If so, the very low residue levels noted for arsenic-treated citrus fruit, instead of presenting a health hazard, might actually be shown to be beneficial.

Control of the Abscission Process for Mechanical Fruit Harvesting. This subject has recently been reviewed by Wilson et al. (51). Although citrus fruit can be successfully harvested without abscission chemicals, field experience has shown that chemical loosening is desirable because less tree shaking time is required, resulting in less physical abuse to the machinery and trees. However, chemicals have not allowed the construction of less-powerful shakers because chemically induced fruit loosening is not always uniform, which results in about 10-15% adhering strongly to the tree.

Although abscission chemicals are technically classified as growth regulators, most function by causing superficial peel burn followed by the fruit producing wound ethylene. The latter moves in some manner through tissue and affects the abscission zone. The only commercially available chemical that appears to function through absorption by tree and fruit, followed by conversion of the chemical into ethylene, is ethephon.

Temperature is the most important physiological factor affecting abscission of early and midseason oranges (52). Prediction of abscission activity is complicated (52), but generally good abscission activity occurs if daily high temperatures are 65°F (18.3°C) or greater following spray application. Daily high temperatures lower than this figure usually result in lessened abscission activity or none at all. Rainfall within 24 hr of a spray application often will negate its effect.

The 'Valencia' orange is harvested in Florida from April until early July and in California, Spain and other Northern Hemisphere citrus-growing areas from April through October. It differs from most fruits because its maturity occurs 13-18 months following bloom, which means that both a mature and immature crop are usually present when the crop is harvested. Cold temperatures are rarely a limiting factor during harvest. However, 'Valencia' undergoes a period of lessened physiological response to abscission chemicals which lasts 2-3 weeks and usually occurs about the first of May (54). During this period the chemical is essentially useless for loosening fruit. Before this period, the fruit is very responsive to abscission chemicals, but the desired Brix/acid ratios (fruit maturity) for processing oranges has not been achieved (55). Following the period, the immature fruit on the tree usually will average 1.5-2.0 cm in diameter and will have achieved sufficient mass so that any mechanical shaking device may remove excessive quantities along with mature fruit.

Five chemicals and chemical combinations are available for use in Florida (51), and a combination of 2 of them has been used successfully. A summary of their uses is as follows:

Ethephon (Ethrel) is cleared by the EPA (Environmental Protection Agency of the United States government) for use in Florida on tangerines and tangerine hybrids (14). This chemical, in addition to producing fruit loosening, enhances fruit color development.

Cycloheximide (Acti-Aid) is cleared by EPA for use in Florida on oranges intended for processing. This chemical has generally produced good loosening of early and midseason oranges, but should not be applied after the spring growth begins, otherwise, severe phytotoxicity can result. Unfortunately, its performance on the 'Valencia' orange has been unacceptable. Cycloheximide should not be used if freezing temperatures are likely to occur because it reduces the cold hardiness of the tree for an undetermined period of time.

5-chloro-3-methyl-4-nitro-1H-pyrazole (Release). This compound is available in Florida for use under experimental permit on oranges destined for processing. The chemical was the first which showed the ability to loosen mature 'Valencia' oranges while causing virtually no injury to bloom, young fruit or foliage when used as recommended.

Glyoxal Dioxime (Pik-Off) is a chemical very similar in mode of action to Release but its experimental use permit has not been renewed.

Chemical Combinations: Two-way combinations of Release and cycloheximide applied with surfactant as dilute sprays have given better fruit loosening than either chemical used alone.

In Australia, abscission tests have generally reported removal force reductions very similar to those noted in Florida (56, 57, 58). Their principal cultivar of interest is the 'Valencia' because the navel orange, although grown extensively in Australia (and to a much lesser extent in Florida) is processed to a very limited extent throughout the world. The tasteless precursor of limonin, namely limonic acid A-ring lactone, is found in raw navel oranges but upon processing, limonin is formed which produces a bitter taste in the products (59). The 'Valencia' orange, early and midseason oranges grown in Florida contain limonin but at relatively low levels.

Minor Commercial Uses and Potentially Rewarding Research Areas

Control of Flowering. The initiation and development of flowers involves a large number of interrelated, well-coordinated growth, senescence and abscission processes (60). The climate in which citrus grows affects its flowering characteristics. In tropical, moist regions where continuous flowering, fruit set and fruit development tend to take place, flower formation usually is relatively light at any one time unless drought conditions intervene to stop growth. In subtropical regions, the advent of spring growing conditions normally produces a single bloom period which may be heavy or light depending on preceding climatic conditions and/or crop load. Some cultivars of lemons, limes and citrons tend to be everbearing (continuously blooming) under all conditions. High temperatures reverse the flowering process and the presence of fruit can inhibit flower formation (61).

A number of chemical compounds promote flowering of certain citrus cultivars (61, 62, 63, 64). As of this date, however, this writer knows of no commercial practice to increase flower formation of citrus by use of plant growth regulators.

Fruit Set. Fruit set in citrus appears to be controlled by the process of competitive inhibition (61). Except for navel, most commercial orange and grapefruit cultivars in the USA set sufficient crops so that the need for a fruit set chemical is not necessary. However, some mandarins and mandarin-grapefruit hybrids (tangelos) benefit from growth regulators which function as fruit-set chemicals. The cultivars 'Orlando', 'Minneola', 'Nova' and 'Robinson' tend to set sparsely but in Florida increased fruit set has been obtained through sprays of GA applied to trees in full bloom (65). Similar results were obtained in South Africa where it is recommended that all bearing blocks of 'Clementine' tangerines be sprayed at full bloom with GA (66). Fruits retained by this method tend to be somewhat smaller and more sensitive to adverse climatic conditions than cross-pollinated fruit (65).

Improved fruit set with navel orange and other cultivars from PGR applications has been reported from many parts of the world (67, 68, 69, 70).

Cold Hardiness. Trees of Citrus sp., although tropical in origin, have the ability to become cold hardened to some extent if subjected to low, but not freezing temperatures, in the presence of light (71). There is no indication, however, that fruit can be cold hardened; hence its protection is restricted to some means of artificial heating or the elimination of ice-initiating bacteria which is currently under investigation (72).

MH (maleic hydrazide) will produce cold-hardening of citrus (9, 71, 73), but in Florida, various bad side effects (delayed regrowth with malformed leaves) have prevented grower acceptance. Under northern California (Mediterranean climate) conditions, night temperatures are sufficiently low during fall that growth protection chemicals are not as important as in areas such as subtropical, moist Florida and subtropical arid Texas and Southern California where "broken winter" conditions exist. This condition is typified by irregular periods of warm or cold conditions which can prevent cold-hardening or cause early breaking of any cold hardiness achieved.

Under Japanese growing conditions (moist, marine), NAA applied in late autumn reduced damage from low temperatures (74). NAA delayed bud break of sweet orange seedlings up to 177 days (75).

Translocation of NAA delayed unsprayed bud growth up to 150 days, but carryover inhibition effect into the second growing season was minimal. Preliminary tests in Florida (3) showed that both NAA and 2,4-DP delayed the growth of young (2-3 year) citrus trees. By summer, however, there were no visible growth differences among any of the treatments.

Control of Tree Growth. In commercial citrus nurseries, stimulating young tree growth would be desirable so that the tree can be brought into production more quickly. Although GA will stimulate growth, it does not appear to accomplish this objective satisfactorily under California conditions (9). However, under some conditions in Florida, it has given significant growth increases (76).

Attempts to use NAA to inhibit rootstock sprouting in nurseries have given mixed results, depending on which rootstock cultivar was used (77).

In Florida grove operations, mature trees of sweet orange, lemon, grapefruit and tangerine cultivars often must be severely hand pruned or machine hedged and topped to prevent tree sizes from exceeding heights of 15-20 ft. (4.6-6.1 meters). In California, many lemons are hand or machine topped yearly to about 10 ft (3 meters) to facilitate harvesting and pest control (78). Although rootstocks to control tree size are being tested in several citrus growing areas, as yet, none are commercially available. Hence, mechanical control mechanisms are necessary; regrowth by citrus is often very rapid and substantial.

Because of the desire to develop a mechanical harvesting system for lemons, considerable tree size control research has been accomplished in both Florida and California (79, 80, 81, 82, 83, 84). All of the treatments, however, appear to have caused some phytotoxicity, although size control was obtained by some of the treatments. A new PGR called P333 (paclobutrazol or (2RS, 3 RS) -1- (4-chlorophenyl) - 4, 4-dimethyl-2 (1, 2, 4-triazol-1-yl) pentan-3-ol (paclobutrazol) has controlled growth on pome fruits (85). Although research with this compound is being conducted on citrus, results have not, as yet, been reported.

Sweet orange, certain tangelos, grapefruit and certain mandarin cultivars could also benefit from size control with PGR's. However, their regrowth problems are generally not as severe as those of lemons.

A white latex paint called Tre-Hold is marketed for control of trunk resprouting (86), and is used extensively in Florida on young citrus trees. It contains 2% ethyl ester of α-NAA.

Color. Two recent publications reviewed color research on citrus (87, 88). The major carotenoid in orange and tangerine peel is B-citraurin. Development of color (B-citraurin) is controlled by an interrelationship between fruit ethylene concentration and temperature (89).

In Florida, preharvest sprays of ethephon are recommended to improve fruit color of mandarins and their hybrids (14). An additional purpose of the sprays is to produce fruit abscission (loosening). The fruit must have achieved minimum internal quality requirements and 10-20% color break before this treatment can be effective (14, 55). No surfactant should be used with ethephon and it should not be combined with other materials containing surfactants; otherwise, increased leaf losses will occur.

Our own observations following ethephon applications to oranges have consistently shown erratic performance and, frequently, heavy leaf losses (3). Similar results seem to have been observed worldwide. Still, its overall advantages outweigh its disadvantages for promotion of color (and abscission) with mandarin-type fruits (14).

Various attempts to control the leaf drop problem caused by ethephon have been made by many researchers, though little of the information seems to have been published, probably because of negative or conflicting results (3).

Another approach to the color problem has been through use of compounds which cause the fruit to produce lycopene rather than B-citraurin (87, 88). The resultant fruit, then, tends to be red and resembles a tomato more than an orange. Best known of these compounds seems to be CPTA [2-(4-chlorophenylthio)-triethylamine hydrochloride] (90). Although, in time, the public could probably be educated to accept oranges which are more "red" than "orange," one of the principal problems with these chemicals is that, under field conditions, they produce an uneven or blotchy color pattern. Our tests have not shown that these compounds increase internal fruit color (3).

Very little has been accomplished concerning improving internal citrus fruit color with PGR applications. Although external peel color is not important for fruit used for processing, internal (juice) color is very important. Oranges grown in the tropics usually have poor external and only fair internal color, but internal qualities such as juice content and soluble solids may be reasonably good (1, 87, 88). However, development of a plant growth regulator to improve internal color, or any of these other qualities, would be beneficial.

Nursery Aids

The cost of a citrus tree can be a substantial part of the initial cost of planting a citrus grove (orchard), hence it would be beneficial to decrease the time of germination of seed, increase the per cent germination and increase the growth rate of the seedling (61). Several methods have been suggested (91, 92, 93, 94, 95, 96, 97).

Plant growth regulators are also reported to aid propagation. The highest number of roots was obtained from air layers of seedless lemon when the stem was treated with an aqueous solution containing a combination of IBA and NAA (98). Cuttings of soft wood or semi-hard wood are reported to respond well to growth regulators such as IBA (61).

Attempts to increase the size, particularly girth, of a seedling using growth regulators have given mixed results. GA applications generally tend to increase the stem length but at the expense of reduced girth (99). However, weekly applications of GA with several spreader-sticker and antitranspirant products significantly increased plant height, intermode length, and stem diameter of sour orange (*C. aurantium* L.) seedlings (76).

Plant growth regulators may be useful for controlling shape of young citrus trees (100, 101). They can also reduce the time lapse between bud grafting and outgrowth of the bud (9, 102, 103).

Conclusion

A great deal of PGR research has been conducted on citrus, and some compounds have achieved commercial importance. Although PGR's continue to have problems relating to their application and use, most appear to be theoretically solvable through continued research programs now in existence. For a more in depth treatment of PGR uses on citrus, see Wilson (3).

Literature Cited

1. Reuther, W. In "Citrus Industry"; Reuther, W. (Editor); Univ. of Calif., Div. Agric. Services, Berkeley, 1973; Vol. 3, p. 280.
2. Wilson, W. C.; Holm, R. E.; Clark, R. K. Proc. Int. Soc. Citriculture, 1977, 2, 404.
3. Wilson, W. C. The Use of Exogenous Plant Growth Regulators on Citrus. In "Plant Growth Regulating Chemicals"; Nickell, L. G. (Editor); CRC Press, Inc., Boca Raton, Florida, 1983; Vol. 1, Chap. 8.
4. Iwahori, S. Proc. Int. Soc. Citriculture, 1978, 263.
5. Hirose, K.; Iwagaki, L.; Suzuki, K. Proc. Int. Soc. Citriculture, 1978, 270.
6. Iwahori, S.; Oohata, J. T. Sci. Hortic., 1976, 4, 167.
7. Wheaton, T. A. Proc. Int. Soc. Citriculture, 1981, 1, 263.
8. Gallasch, P. T. Proc. Int. Soc. Citriculture, 1978, 276.
9. Coggins, C. W., Jr.; Hield, H. Z. In "The Citrus Industry"; Revised ed., Reuther, W.; Webber, H. J.; Batchelor, L. D. (Editors); Univ. of Calif., Div. Agric. Services, Berkeley, 1968, Vol. 1, p. 371.
10. Monselise, S. P. Sci. Hortic., 1979, 11, 151.
11. Gallasch, P. T.; Bevington, K. B.; Hocking, D.; Moss, G. I. Proc. Int. Soc. Citriculture, 1978, 273.
12. Moss, G. I.; Bevington, K. B. Proc. Int. Soc. Citriculture, 1978, 260.
13. Moss, G. Rural Research, 1977, 94, 22.
14. Knapp, J. L. Fla. Coop. Ext. Ser., Circular 393-I (revised annually), 1983.
15. Stewart, W. S.; Hield, H. Z. Proc. Am. Soc. Hortic. Sci., 1950, 55, 163.
16. Sarooshi, R. A.; Stannard, M. C. Aust. J. Exp. Agric. Anim. Husb., 1975, 15, 429.
17. Krezdorn, A. H. Calif. Citrogr., 1977, 62, 85.
18. Dinar, H. M. A.; Krezdorn, A. H.; Rose, A. J. Proc. Fla. State Hortic. Soc., 1976, 89, 4.
19. El-Zeftawi, B. M. J. Hortic. Sci, 1980, 55, 211.
20. Gilfillan, I. M.; Koekemoer, W.; Stevenson, J. The Citrus Grower and Sub-Tropical Fruit Journal, 1972, 458, 6.
21. Gilfillan, I. M.; Koekemoer, W.; Stevenson, J. Proc. Int. Soc. Citriculture, 1973, 3, 335.
22. Gilfillan, I. M.; Stevenson, J. A. Citrus Sub-Trop. Fruit J., 1976, 507, 5.
23. Ramirez, J. M.; Krezdorn, A. H.; Rose, A. J. Proc. Fla. State Hortic. Soc., 1977, 90, 61.
24. Grierson, W. Proc. Int. Soc Citriculture, 1981, 2, 764.

25. Monselise, S. P. Proc. Int. Soc. Citriculture, 1978, 250.
26. Monselise, S. P.; Weiser, M.; Shafir, N. ; Goren, R.; Goldschmidt, E. E. J. Hortic. Sci., 1976, 51, 341.
27. Gilfillan, I. M.; Stevenson, J. A.; Wahl, J. P. Proc. Int. Soc. Citriculture, 1981, 1, 224.
28. Gilfillan, I. M.; Stevenson, J. A.; Holmden, E.; Ferreira, C. J.; Lee, A. The Citrus and Subtrop. Fruit J., 1980, 605, 11.
29. Kuraoka, T.; Iwasaki, K.; Ishii, T. J. Am. Soc. Hortic. Sci. 1977, 102, 651.
30. Gilfillan, I. M. "Report on a visit to Japan, Australia, and the United States of America with Emphasis on Fruit Size Improvement," Outspan Citrus Center, Nelspruit, South Africa, 1983.
31. Burns, R. M., 1982, (personal communication).
32. El-Zeftawi, B. M. Sci. Hortic., 1980, 12, 177.
33. Grierson, W.; Hatton, T. T. Proc. Int. Soc. Citriculture, 1977, 1, 227.
34. Wardowski, W. F.; Barmore, C. R.; Smith, T. S.; DuBois, C. W. Proc. Fla. State Hortic. Soc., 1974, 87, 216.
35. Jahn, O. L. Proc. Fla. State Hortic. Soc., 1974, 87, 218.
36. Brown, G. E.; Barmore, C. R. Proc. Fla. State Hortic. Soc., 1978, 89, 198.
37. Gilfillan, I. M., 1982 (personal communication).
38. Bleinroth, E. W.; Hansen, H. A.; Ferreira, V. L. P.; Angelucci, E. Coletanea do Instituto de Technologia de Alimentos, 1976, 7, 343.
39. Ismail, M. A.; Grierson, W. HortScience, 1977, 12, 118.
40. Feinstein, B.; Monselise, S. P.; Goren, R. HortScience, 1975, 10, 385.
41. Lewin, I. J.; Monselise, S. P. Sci. Hortic., 1976, 4, 229.
42. Ferguson, L.; Davies, F. S.; Ismail, M. A.; Wheaton, T. A. Plant Growth Reg. Soc. of Am., 1983, 10, 175.
43. Wilson, W. C.; Kenny, D. S.; Holm, R. E. Proc. Int. Soc. Citriculture, 1977, 2, 692.
44. Crous, P. A., Citrus Grower, 1941, 94, 1.
45. Mynhardt, C. B. Citrus Grower, 1956, 273, 1.
46. Basson, W. J. Farming in South Africa, 1959, July, p. 52.
47. Gilfillan, I. M.; Stevenson, J. A. The Citrus & Subtrop. Fruit J., 1976, March, p. 13.
48. Gilfillan, I. M.; Wahl, P.; Holmden, E.; Reay, N.; Stevenson, J. A. The Citrus & Subtrop. Fruit J., 1975, March, p. 5.
49. Kesterson, J. W.; Braddock, R. J.; Koo, R. C. J.; Reese, R. L., HortScience, 1975, 10, 65.
50. Metz, W. Science, 1981, 213, 1332.
51. Wilson, W. C.; Coppock, G. E.; Attaway, J. A. Proc. Int. Soc. Citriculture, 1981, 1, 278.
52. Biggs, R. H.; Kossuth, S. V. Acta Hortic., 1981, 120, 71.
53. Wilson, W. C. Proc. Plant Growth Reg. Soc. of Am., 1980, 7, 41.
54. Wheaton, T. A.; Wilson, W. C.; Holm, R. E. J. Am. Soc. Hortic. Sci., 1977, 102, 580.
55. Wardowski, W.; Soule, J.; Grierson, W.; Westbrook, G. Fla. Cooperative Extension Services, Univ. of Fla., Bull. 188, 1979.
56. Hutton, R. J. Proc. Int. Soc. Citriculture, 1978, 257.

57. Hutton, R. J. Proc. Int. Soc. Citriculture, 1981, 1, 281.
58. El-Zeftawi, B. M. Proc. Int. Soc. Citriculture, 1978, 255.
59. Ting, S. V.; Fisher, J. F.; Nagy, S. , 1982, (personal communication).
60. Goldschmidt, E. E. Plant & Cell Physiol., 1980, 21, 193.
61. Moss, G. I. Ciba-Geigy Agrochemicals Tech. Monograph, 1975, No. 4, p. 61.
62. Nir, I.; Goren, R.; Leshem, B. J. Am. Soc. Hortic. Sci., 1972, 97, 695.
63. Goren, R.; Monselise, S. P. Planta (Berl.), 1970, 88, 364.
64. Salomon, E. Acta Hortic., 1980, 114, 11.
65. Krezdorn, A. H.; Jernberg, D. C. Proc. Int. Soc. Citriculture, 1977, 2, 660.
66. de Lange, J. H.; du Plessis, S. F.; Vincent, A. P.; du Preez, M. B.; Holmden, E. A.; Rabe, E. Subtropica, 1982, 3, 7.
67. Dayuan, Wang. HortScience, 1981, 16, 657.
68. Shawki, I.; El-Tomi, A.; Nasr, A. Egypt. J. Hortic., 1978, 5, 115.
69. Kumar, R.; Singh, J. P.; Gupta, O. P. Haryana J. Hortic. Soc., 1975, 4, 123.
70. Garcia-Martinex, J. L.; Garcia-Papi, M. A. Scientia Hortic., 1979, 10, 285.
71. Yelenosky, G. Proc. Int. Soc. Citriculture, 1977, 1, 199.
72. Lindow, S. E.; Arny, P. C.; Upper, C. D.; Barchet, W. R. In "Plant Cold Hardiness and Freezing Stress," Li, P. H.; Sakai, A. (Editors), Academic Press, New York, 1978, p. 249.
73. Hendershott, C. H. Proc. Am. Soc. Hortic. Sci, 1962, 80, 241.
74. Konakahara, M. Special Bulletin of the Shizuoka Prefectural Citrus Experiment Station No. 3, Komagoe, Shimizu-City, Japan, 1975, 1.
75. Nauer, E. M.; Boswell, S. B.; Holmes, R. C. HortScience, 1979, 14, 525.
76. Miller, I. A.; Young, M. J. HortScience, 1982, 17, 673.
77. Nauer, E. M.; Boswell, S. B. HortScience, 1978, 13, 166.
78. Burns, R. M., 1983, (personal communication).
79. Lundberg, E. C.; Smith, T. S. Proc. Fla. State Hortic. Soc., 1974, 87, 20.
80. Moye, H. A.; Willson, A. E. Proc. Fla. State Hortic. Soc., 1977, 90, 272.
81. Phillips, R. L.; Tucker, D. P. H. HortScience, 1974, 9, 199.
82. Boswell, S. B.; McCarty, C. D.; Ede, L. L.; Chesson, J. H. Citrograph, 1975, 60, 405.
83. Boswell, S. B.; Burns, R. M.; Hield, H. Z. HortScience, 1976, 11, 115.
84. Boswell, S. B.; Embleton, T. W.; Jones, W. W.; Summers, L. L. HortScience, 1981, 16, 41.
85. Williams, M. W. Plant Growth Reg. Soc. of Am., 1982, 9, 78.
86. Plant Growth Regulator Handbook, 2nd Edition. Plant Growth Reg. Soc. of Am., Lake Alfred, Fl, 1981, p. 78.
87. Stewart, I. Proc. Int. Soc. Citriculture, 1977, 1, 308.
88. Stewart, I. In "Citrus Nutrition and Quality," Nagy, S.; Attaway, J. A. (Editors); ACS Symposium Series 143; American Chemical Society, 1980, p. 129.
89. Wheaton, T. A; Stewart, I. J. Am. Soc. Hortic. Sci., 1973, 98, 337.

90. Yokoyama, H.; Hsu, W.; Poling, S. M.; Hayman, E.; De Benedict, C. Proc. Int. Soc. Citriculture, 1977, 3, 717.
91. Burns, R. M.; Coggins, C. W. Calif. Agric., 1969, 23, 18.
92. Choudhari, B. K.; Chakrawar, V. R. J. Maharashtra Agric. Universities, 1980, 5, 173.
93. Choudhari, B. K.; Chakrawar, V. R. J. Maharashtra Agric. Universities, 1981, 6, 19.
94. Shukla, K. S.; Misra, R. L.; Kaul, M. K.; Prasad, A. Haryana J. Hortic. Sci., 1978, 7, 162.
95. Abdalla, K. M.; Wakeel, A. T. El; Masiry, H. H. El. Research Bulletin, Ain Shams Univ., Faculty of Agric. Zagazig Univ., Cairo, Egypt, 1978, No. 944.
96. Misra, R. S.; Verma, V. K. Progressive Hortic., 1980, 12, 79.
97. Singh, H. K.; Shankar, G.; Makhija, M. Haryana J. Hortic. Sci., 1979, 8, 194.
98. Patil, S. B.; Chakrawar, V. R. Punjab Hortic. J., 1979, 19, 110.
99. Abdalla, K. M.; El-Wakeel, A. T.; E.-Masiry, H. H. Research Bulletin, Faculty of Agric., Ain Shams Univ., Zagazig Univ., Zagazig, Egypt, 1979, No. 937.
100. Abdalla, K. M.; Wakeel, A. T. El; Masiry, H. H. El. Research Bulletin, Ain Shams Univ., Zagazig Univ., Cairo, Egypt, 1978, No. 936.
101. Hassaballa, I. A. Research Bulletin, Ain Shams Univ., Faculty of Agric., Zagazig Univ., Moshtohor, Egypt, 1979, No. 1076.
102. Maiti, R. B.; Singh, S. M.; Singh, I. J. Indian J. Hortic., 1959, 16, 249.
103. Mauer, E. M.; Boswell, S. B.; Holmes, R. C. HortScience, 1979, 14, 229.

RECEIVED February 22, 1984

[14C]Abscisic Acid Transport and Metabolism in Source and Sink Tissues of *Beta vulgaris*

JALEH DAIE and ROGER WYSE

Department of Biology, Utah State University, and Crops Research Laboratory, Agricultural Research Service, U.S. Department of Agriculture, Logan, UT 84322

The mode of abscisic acid (ABA) uptake and its metabolism was studied in excised leaf and root discs of sugarbeet (Beta vulgaris L.). Uptake by both tissues was highly pH-dependent. Maximal uptake was observed at pH 5.5. ABA moves across membranes as the protonated lipophilic species. As a weak acid, ABA dissociates and accumulates in the more alkaline compartment. The distribution of ABA within the tissue is regulated by the pH differential between any two compartments. Although diffusion may be the predominant form of transport, the uptake of ABA is dependent on metabolic energy for the establishment of a pH gradient across the membrane.

Dihydrophaseic acid (DPA), DPA aldopyranoside, phaseic acid (PA), ABA glucose ester (ABA-GE), and at least two unidentified compounds were recovered from both tissues. An additional unidentified metabolite was present only in root tissue. Both tissues used the same metabolic pathway. However, the pathways were utilized at different rates by the two tissues. Leaf discs exhibited a higher capacity for ABA conjugation, and root discs showed a greater preference for ABA catabolism to PA and DPA. After 13 h, the major metabolites in the leaf were the ABA-GE, DPA, and PA. In the root the predominant metabolites were PA and the unique, unknown metabolite. Internal ABA pool size was independent of external concentrations that were not more than 10^{-6} M, suggesting that rates of ABA metabolism were proportional to the rates of uptake in both tissues. These data suggested that both source and sink tissues can effectively metabolize ABA by conjugation or catabolism. We propose that over-all endogenous ABA levels among different plant parts are more likely to be the result of relative rates of synthesis vs. metabolism, while a pH differential across membranes regulates subcellular ABA concentrations.

0097–6156/84/0257–0127$06.00/0

The principles governing distribution, transport, and metabolism of ABA are not well understood. An understanding of these principles is important because it provides information about factors influencing hormone concentrations at sites of action and may eventually explain the unusually high concentration of ABA found in some organs or tissues.

ABA is synthesized predominately in the leaves and is transported in the phloem along with other assimilates and is redistributed through various tissues and cells where it exerts its effects on different physiological processes. Recently ABA has been implicated in the control of photosynthate partitioning and in source-sink communication (1, 2). The role, if any, for ABA in sink metabolism is unclear but it may influence phloem unloading (1) and sucrose uptake (3). However, if ABA is continuously transported to sink regions, it must either accumulate to high levels or be readily metabolized. Studies of ABA metabolism are important, therefore, because they help clarify whether an observed effect was due to ABA or to a catabolic product made by the plant when ABA was applied exogenously. Similarly, without information on ABA metabolism, a lack of any ABA effect may be an erroneous interpretation due to possible ABA inactivation (conjugation or degradation).

We have observed stimulation of sucrose uptake in sugarbeet root tissue (Figure 1). However the results were erratic. Therefore we attempted to answer the following questions: 1) Is exogenous ABA taken up by the tissue? 2) If yes, is exogenously applied ABA stable in the tissue? and 3) How do physiologically different tissues (source-sink) differ in their ABA transport and metabolism?

Materials and Methods

Plant Material. Seeds of sugarbeet (*Beta vulgaris* L. cv AH-11) were grown under greenhouse conditions. Fully expanded mature leaves and roots of 5-month-old plants were collected immediately prior to any experiment.

Preparation of Tissue Discs. Leaf discs 10 mm in diameter were obtained using a sharp corkborer. Discs from several leaves were pooled and used within 10 min without pretreatment. Root tissue was prepared by cutting slices 1 mm thick with a hand microtome. Discs 5 mm in diameter were cut from these slices with a sharp corkborer and equilibrated for 90 min in a solution containing 1 mM $CaCl_2$ and 200 mM mannitol buffered at pH 6.5 with 30 mM MOPS-BTP (Morpholinopropanesulfonic Acid-Bis Tris Propane). The equilibration medium was aerated continuously and replaced every 15 min.

Incubation. Unless otherwise stated, the base incubation medium for leaves contained 10 mM MES-BTP (N-morpholine ethanesulfonic acid - Bis Tris Propane) (pH 5.5), 1 mM $CaCl_2$, and 1 M [^{14}C]ABA (10^5 dpm/rep). Five leaf discs (0.1 g fresh weight) were floated (upper side up) on 3 ml of medium in a 4-cm diameter petri dish. The incubation medium for root tissue consisted of 30 mM MES-BTP (pH 5.5), 1 mM $CaCl_2$, 1 M [^{14}C]ABA (10^5 dpm/rep), and 200 mM mannitol as the osmoticum. Thirty discs of root tissue (0.75 g fresh weight) were incubated in 3 ml of medium in a 20-ml vial. All treatments were run in triplicate, and all experiments were repeated at least twice. For more details refer to Daie and Wyse (4).

Metabolite Extraction and Analysis. Immediately after the washout procedure, tissues were frozen in liquid N_2, freeze-dried, and homogenized in chilled 90% MeOH (20 ml/g fresh weight). After filtration, the extracts were taken to dryness under reduced pressure at 25 C and redissolved in 1.5 ml of 0.2 N acetic acid. Particulates were removed by filtration through a 1.2 µm nitrocellulose filter. The filtrate was then chromatographed by reverse-phase HPLC (High Performance Liquid Chromatography). The preparative HPLC column was a 150 X 10 mm C_{18} (RSIL, 10µm particle size). A two-step linear solvent gradient delivered at 2 ml/min was used to separate the metabolites. For details refer to Daie *et al.* (5).

Results

Effect of Time and Concentration. After 30 min, ABA uptake was linear for 5 h by root and leaf tissue (Figure 2). Since uptake varied with time, only total uptake per g fresh weight of tissue is reported. ABA uptake was proportional to external ABA concentrations between 1 nM and 1 µM in both tissues and, although uptake rates declined above 1 µM, there was no evidence of saturation kinetics (Figure 3). At all concentrations, uptake rates were higher by root than by leaf tissue. The greater total uptake in roots may be due to the absence of a cuticular restriction.

Metabolic Inhibitors. To determine if the uptake required metabolic energy, metabolic inhibitors such as CCCP (carbonyl cyanide m-chlorophenyl hydrazone), FCCP (carbonyl cyanide p-trifluoromethoxyphenyl hydrazone), NaN_3, NaCN, and DNP (dinitrophenol) were used. All, except NaCN in root tissue, significantly inhibited ABA uptake (Table I).

Table I. Inhibition of ABA Uptake by Metabolic Inhibitors

Treatment	Inhibitor Conc.	Leaf	Root
	M	pmol g^{-1}	
Control		600 ± 21	1,100 ± 34
CCCP	5×10^{-6}	341 ± 62	721 ± 27
FCCP	5×10^{-6}	352 ± 56	724 ± 34
NaN_3	1×10^{-4}	321 ± 69	841 ± 6
NaCN	1×10^{-4}	498 ± 54	1,039 ± 93
DNP	2×10^{-3}	212 ± 24	514 ± 15

Tissues were incubated at 1 µM ABA (pH 5.5) and the inhibitors for 5 h.

All inhibitors were more effective in leaf tissue. The partial inhibition suggested that the ABA uptake was a combination of energy-dependent and passive transport. CCCP was used in subsequent studies to separate energy-dependent and passive uptake. CCCP was chosen because its activity is not pH dependent. It is a proton

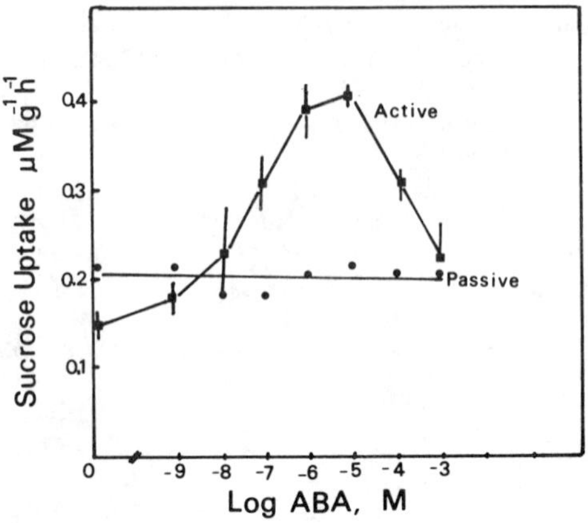

Fig. 1. Effect of ABA on sucrose uptake by sugarbeet root tissue discs.

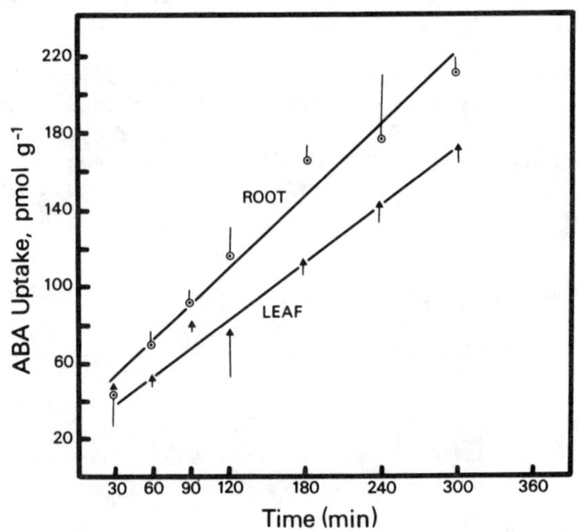

Fig. 2. Time course of ABA uptake. Tissues were incubated in 1 µM [^{14}C]-ABA (pH 5.5). "Reproduced with permission from Ref. 4. Copyright1983, Amer. Soc. Plant Physiol."

ionophore and a metabolic uncoupler. We have previously shown 5 μM CCCP to inhibit respiration strongly in sugarbeet discs (3).

Temperature Effect. To test for the presence of energy-dependent transport by another method, ABA uptake was measured at 17 or 27 C Uptake was higher by both tissues at 27 C (Table II).

Table II. Effect of Temperature on ABA Uptake

Treatment	ABA Uptake		Q_{10}	
	27° C	17° C	Passive	Active*
	pmol			
Leaf				
Control	355 ± 6	195 ± 12	1.82	
CCCP, 5 μM	315 ± 25	150 ± 16	1.43	3.1
DNP, 0.2 mM	135 ± 61	90 ± 30	1.50	2.0
Root				
Control	1,000 ± 29	500 ± 11	2.00	
CCCP, 5 μM	240 ± 8	150 ± 16	1.60	2.2
DNP, 0.2 mM	240 ± 13	165 ± 20	1.50	2.3
CCCP + DNP	245 ± 2	185 ± 2	1.30	2.4

*Energy dependent.
Tissue discs were incubated for 5 h in 1 μM ABA medium.

In the absence of the inhibitor, the Q_{10} for leaves was 1.82 and 2.0 for root tissue. The magnitude of these values did not show conclusively the dominance of either energy-dependent or passive uptake.

Effect of pH. In both tissues, ABA uptake decreased as the pH of the medium was increased (Figure 4). Maximum observed uptake was at pH 5.5. At pH 7.5 or higher, almost no uptake was observed. At pH 5.5, leaves had more energy-dependent uptake than passive uptake. While in roots, passive uptake was greater than energy-dependent uptake. The passive uptake of leaf tissue was independent of external pH. Passive uptake in the root was highly pH dependent and was comparable to energy-dependent uptake in that tissue. At pH 5.5, total uptake in the root tissue was about twice that in the leaf. This could be the result of a physiological difference between root and leaf or because root tissue was more exposed to the medium; i.e.: cut root tissue was submerged in the medium, but leaves were floating on it. Higher ABA uptake at lower pH values suggested that, as a weak acid (pKa = 4.6), ABA was most permeable in its undissociated, lipophilic form.

Light Effect. Leaf discs were incubated in total darkness or under a combination of sodium vapor and incandescent lamps (800 μE $m^{-2}s^{-1}$) during a 5 h of incubation. Total and passive ABA uptake in darkness was about 50% of that in light (Table III). Higher ABA uptake under light conditions did not, by itself, support the conclusion that the uptake was an energy-dependent process because the energy-dependent component (CCCP-sensitive) was not affected. Rather the data suggested that light influences some factor involved in passive uptake. Assuming that the compartments with high pH accumulate ABA, this light effect may be explained by the following:

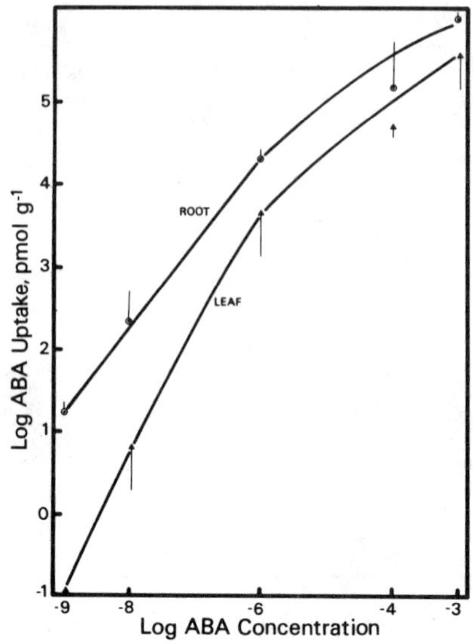

Fig. 3. Effect of external concentration on ABA uptake. [^{3}H] ABA was used at 10 and 1 nM and [^{14}C]ABA at higher concentrations. "Reproduced with permission from Ref. 4. Copyright 1983,Amer. Soc. Plant Physiol."

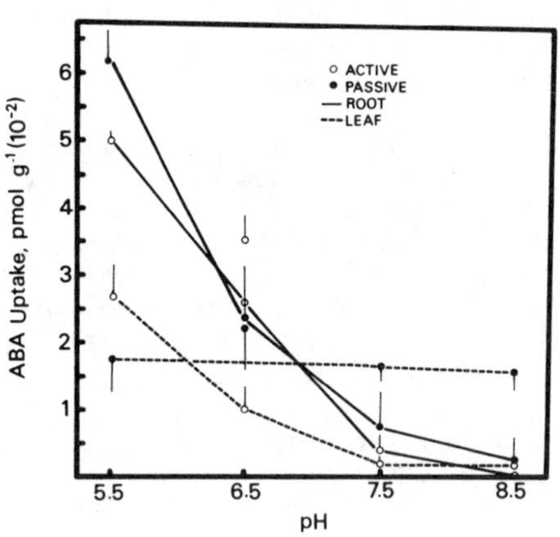

Fig. 4. pH dependence of ABA uptake. "Reproduced with permission from Ref. 4. Copyright 1983, Amer. Soc. Plant Physiol."

Table III. Effect of Light on ABA Uptake

Condition	ABA Uptake	
	Passive	Active
	pmol g^{-1}	
Light	255 ± 13	136 ± 28
Dark	97 ± 7	135 ± 52

Leaf discs were incubated for 5 h in covered containers in a growth chamber at 26° C with a photon flux density of 800 $\mu E \cdot m^{-2} \cdot s^{-1}$. Dark controls were wrapped in foil with allowance for air exchange.

in light, the pH of the chloroplast stroma becomes more alkaline than the cytoplasmic pH; therefore, ABA accumulates inside the chloroplast. This additional compartmentalization would favor greater uptake of ABA.

Metabolites. Analysis of root and leaf tissue extracts for radiolabeled metabolites indicated that labeled ABA was metabolized into several different products. According to their retention times, these metabolites were tentatively identified to be DPA aldopyranoside, DPA, PA, ABA-GE, ABA, $R_{20.5}$ and $R_{27.5}$. The latter two metabolites were unidentified metabolites present in both tissues. An additional unidentified metabolite with a retention time of 25 min (R_{25}) was unique to the root extract.

Effect of Time on Metabolism. After an approximately 2-h lag, labeled ABA uptake by leaf was primarily conjugated to the ABA-GE or catabolized to DPA and PA (Figure 5). After this initial 2-h lag, the incorporation of radioactivity into these metabolites was essentially linear. At the end of the 13-h incubation period, leaf discs had converted 60% more ABA into the ABA-GE than had been accumulated in the PA fraction. However, the amount of label in the combined PA and DPA pool was about the same as that of ABA-GE.

In the root, labeled ABA was rapidly incorporated into PA and the unique metabolite (R_{25}) with no apparent lag period (Figure 5). The ABA-GE accumulated in the root tissue at much slower rates than did PA. In contrast to the leaf, after 13 h, root tissue had incorporated almost twice as much ABA into PA as into ABA-GE.

Concentration Effect on Metabolites. By measuring total dpm in the tissue, Daie and Wyse (4) had previously shown that an increase in the exogenous ABA concentration from 10^{-9} M to 10^{-3} M resulted in a proportional increase in ABA uptake. ABA metabolism, indicated by incorporation of label into different ABA metabolites, was also proportional to exogenous ABA concentrations (Figure 6). Substantantially larger amounts of ABA were converted into its metabolites when tissue was incubated in the presence of 10^{-5} M ABA than with 10^{-8} M ABA.

Internal ABA Pool. To determine how the internal ABA pool was affected as a result of incubation with external ABA, tissues were incubated for 5 h in the presence or absence of 10^{-6} M unlabeled

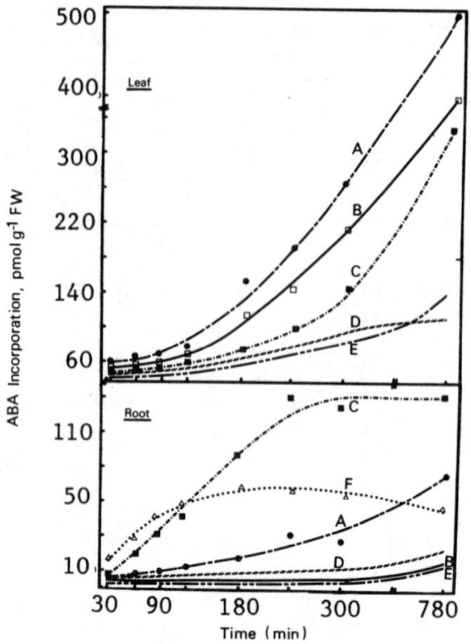

Fig. 5. Time course study of [^{3}H] metabolism by sugarbeet leaf and root tissue discs. Samples were taken at all time points shown on the horizontal axis. ABA-GE (A), DPA (B), PA (C), $R_{27.8}$ (D),DPA aldopyranoside (E), and R_{25} (F).

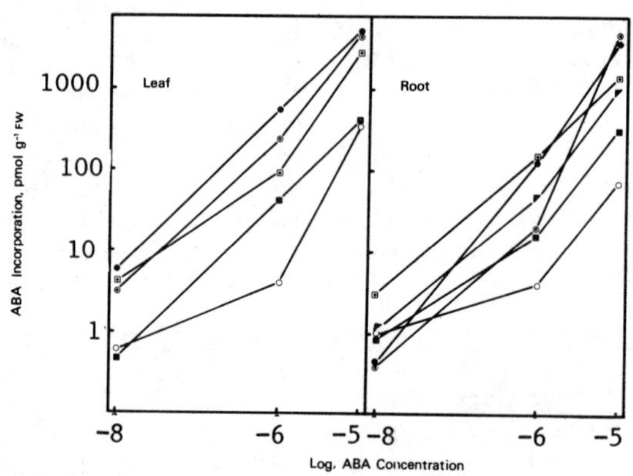

Fig. 6. Effect of exogenous ABA concentrations on [^{3}H]ABA incorporation into metabolites. Tissue was incubated for 5 h in media as described in Materials and Methods. (ABA (●), ABA glucose ester (⊙), PA (⊡), $R_{27.8}$ (■), DPA aldopyranoside (○), and the unidentified metabolite (R_{25}) of the root extract (▲).

ABA. Endogenous ABA concentrations were not different in the presence or absence of unlabeled ABA in the medium (Table IV).

Table IV. Endogenous ABA Levels of Leaf and Root Tissue Incubated for 5 h with or without 10^{-6} M Unlabeled ABA

	Time (min)			
Leaf	0		20 ± 10	
	150	28 ± 8		30 ± 4
	300	15 ± 3		16 ± 4
Root	0		22 ± 1	
	150	25 ± 5		26 ± 3
	300	19 ± 2		33 ±10

Date represent mean of 4 determinations ± SE.

Considering the previously defined rates of ABA uptake (4), and if exogenous ABA were not significantly metabolized, we expected to see increases of up to 3-fold in endogenous ABA levels during a 5-h incubation. To further test for the presence of a rapid mechanism for control of ABA levels, tissues were incubated with 100-fold higher external ABA concentration (10^{-4} M) than used in the previous experiment. After 5 h, tissue incubated with 10^{-4} M ABA had significantly more endogenous ABA than did the tissue incubated with 0 or 10^{-6} M ABA (Table V). This experiment also indicated that no

Table V. Endogenous ABA Levels in Leaf and Root Tissues Incubated for 5 h with Various Concentrations of Unlabeled ABA

		Applied Concentration	
	Time (min)	Applied (M)	Endogenous (ng g^{-1} fw)
Leaf	0	0	16 ± 4
	300	0	14 ± 11
		10^{-6}	55 ± 27
		10^{-4}	640 ± 42
Root	0	0	7 ± 1
	300	0	5 ± 2
		10^{-6}	15 ± 9
		10^{-4}	1630 ± 80

stress-induced ABA was produced by the tissues during incubation since, in the absence of any exogenous ABA, the endogenous ABA levels were the same at 0 time and after 5 h (Table V). The data suggested a relatively constant internal pool of ABA despite continued ABA uptake. The increase in endogenous ABA when tissues were incubated with higher exogenous ABA concentration than 10^{-6} M, indicated the inability of the tissue to metabolize ABA at rates commensurate with high rates of uptake.

The fact that endogenous ABA levels reach steady state conditions after about 4 h of incubation is illustrated in Figure 7. The total amount of radioactivity present in the tissue continued to rise in both tissues, but after 3-4 h, the amount of radioactivity found in the ABA fraction stabilized. During the last 9-10 h of incubation, the ABA pool was in a steady state condition, and the specific activity remained constant.

Discussion

Use of the uncoupler and proton ionophore CCCP has been questioned by some investigators because it has been associated with membrane damage in some tissues (6). We used leakiness of the membranes to sucrose as an index of membrane damage. Inasmuch as sugarbeet root cells store a high concentration of sucrose in the vacuole, any membrane damage should cause leakage of sucrose into the incubation medium. The sucrose content of the medium after 5 h of incubation in the presence of 5 μM CCCP was not different from the medium without CCCP (data not shown). Thus 5 μM CCCP had no apparent effect on membrane integrity.

Our present data are consistent with previous reports that ABA moves due to a pH gradient across membranes as the undissociated lipophilic species and accumulates as the dissociated species in the more alkaline compartments (7, 8, 9). The very low ABA content of acidic vacuoles of citrus fruit vesicles compared to the ABA content of the cytoplasm can be explained as a pH gradient effect (10). Heilmann (7) calculated that, under normal conditions, 70 to 80% of the ABA in leaf cells is found in the chloroplasts. He attributed this accumulation to a pH differential between cytoplasm and chloroplast (at stroma pH in the range of 7.0-8.0, 95% of the ABA is dissociated and, therefore, is trapped inside the chloroplast). Noriaki et al. (9) calculated that ABA was 80%, 25%, and 4% undissociated at pH's 4, 5, and 6 respectively. They suggested that the increased ABA uptake was caused by diffusion rather than by carrier-mediated transport suggested by Astle and Rubery (11). These latter authors have proposed that a carrier-mediated transport system is partially responsible for total ABA uptake.

Heilmann et al. (7) assume that the tonoplast and plasmalemma membranes would have the same permeability characteristics as the chloroplast envelope. Therefore, pH would be important in a cell-to-cell movement as well as within a cell.

Although several investigators have concluded that ABA is taken up solely via diffusion (7, 8), Astle and Rubery (11) observed a saturating uptake component in roots of runner bean, along with a diffusion component, and they proposed the existence of a carrier-mediated uptake system.

Using isolated mesophyl cells, Kaiser and Hartung (8) recently concluded that ABA transport occurs by a simple diffusion process depending solely on the pH gradient between two compartments or cells. This conclusion would explain our results. CCCP, a proton ionophore and metabolic uncoupler, would dissipate any pH gradient across the membrane. Therefore, uptake in the presence of CCCP would be passive diffusion driven only by the chemical potential gradient of undissociated ABA. However, the establishment of a pH gradient is an active process utilizing ATP and a vectorial ATPase. The net result of this active process is ABA transport. Although movement of ABA is passive *per se*, the accumulation of ABA is linked to the electrogenic process of proton pumping and is thus indirectly energy-dependent.

Transport of ABA due to a pH differential across the membrane seems to be important at localized areas, such as between two compartments of a cell. Establishment of a pH gradient explains the energy-dependent uptake of undissociated ABA. Nevertheless, at physiological pH of 6 to 7, more than 90% of the ABA is in the ionic form. Therefore, it is conceivable that plants may have a mechanism whereby movement of ionic ABA is directly coupled to metabolic energy.

Our data confirmed the existence in both source and sink tissues of sugarbeet of the two metabolic pathways reported for ABA inactivation; namely, conjugation and catabolism to PA and DPA (Figure 8) (12). Although ABA was conjugated in both root and leaf, ABA and DPA were conjugated (ABA-GE and DPA aldopyranoside) more rapidly in the leaf; possibly because of a more efficient compartmentalization mechanism or the presence of more active enzymes in leaf tissue. Another speculation as to why sink tissue would preferentially catabolize ABA as opposed to conjugating it, may be the importance of sugar in sink vs. source tissue. While glucose may be readily available in source tissue for different processes, in sink tissue its use in structure or storage carbohydrates may result in its limited availability. In leaves of *Xanthium*, ABA is efficiently and rapidly conjugated to its glucose ester and is presumably stored without further metabolism (13). This observation on the conversion of ABA to its glucosyl ester in *Xanthium* plants, even after they are released from wilt conditions, is especially interesting because the plant can thus dispose of large quantities of ABA by converting it into a compound with little or no biological activity. The compartmentalization of the ABA-GE with no further metabolism, implies *de novo* synthesis of ABA (rather than release from the bound form) under most conditions (12, 13, 14, 15). Although Zeevaart (16) recently reported that the ABA-GE levels of *Xanthium* remain unchanged 34 days after release from stress, there is no conclusive evidence that further metabolism of the ABA-GE does not occur over the long-term. These data collectively suggest the central role of metabolism and synthesis in the regulation of endogenous ABA levels.

The relative activity of the two pathways in the two tissues studied can be estimated by comparing the relative amounts of labeled ABA converted into end products of either the PA or ABA conjugation pathways. In leaves, labeled ABA was converted to its glucosyl ester at much faster rates and with a shorter lag time than it was incorporated into PA. However, assuming that all labeled DPA was the result of PA breakdown, the activity of each pathway would be about equal in leaves. The lag in the formation of PA and DPA indicated slow initial rates, possibly due to limited substrate. In contrast, root tissue clearly showed much more activity in the PA pathway than in the conjugation pathway. Also the 1-2 h lag in leaf tissue was not observed in root tissue. Another difference between the two tissues seemed to be the reduction of PA to DPA--rapid in leaves but very slow in roots (Figure 5).

Since synthetic (±)-ABA was used in these studies, we do not know how the data would differ if only the naturally occurring (+)-ABA enantiomer had been utilized. It has been suggested that the (-)-ABA enantiomer is mainly derivatized to the ABA-GE (10).

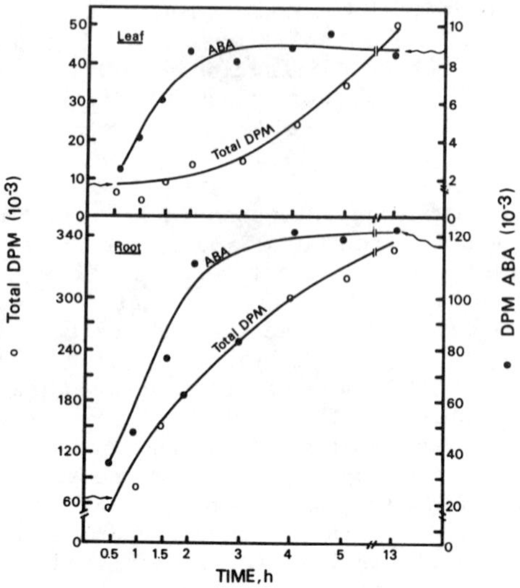

Fig. 7. Relationship between total radioactivty present in the tissue and the radioactivity in the ABA fraction.

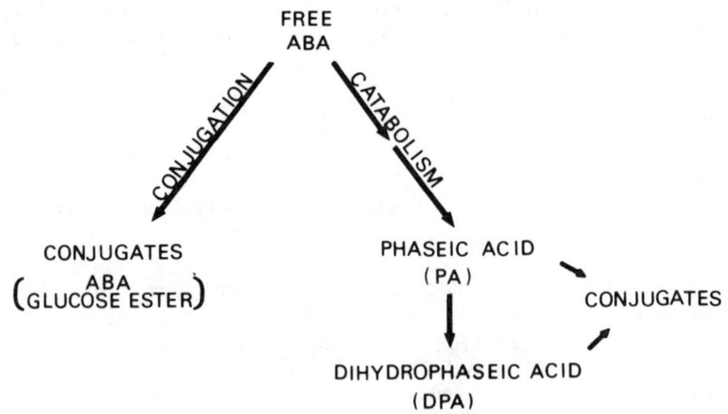

Fig. 8. The two proposed metabolic pathways for ABA.

In the present study, the observed high incorporation of radioactivity into the ABA-GE by leaf tissue may have been a contribution of the (-)-ABA enantiomer. However root tissue received the same ratio of (-)-ABA and yet contained much smaller quantities of ABA-GE than did leaf tissue. We were unable to calculate the differences in rates of conversion of ABA to different metabolites caused by the presence of (-)-ABA. If we assume a 50:50 mixture of the two enantiomers, the in vivo rates for conversion of ABA into its glucosyl ester may be half of the observed in vitro rates in either tissue.

The rise in labeled metabolite concentrations that was concurrent with increases in exogenous ABA concentrations and ABA uptake rates suggested that the plant has the capacity to rapidly metabolize ABA to modify endogenous ABA levels. This conclusion seems logical since unusually high ABA levels are known to be synthesized in a relatively short period of time to trigger a specific response in the plant; i.e., stomatal closure (17). Such high concentrations do not seem to be necessary during the post-reponse period and, therefore, must be disposed of by endogenous mechanisms.

As other investigators have also noted (12, 16), our data are consistent with the hypothesis that plants have a substantial capacity to degrade ABA. This capacity has been substantiated in experiments where plants were subjected to excessive, stress-induced, internal concentrations of ABA (16, 18). This degradative capacity seems to be operative when ABA is applied exogenously. Since the ABA status of the tissue depends on its rates of uptake and metabolism of exogenous ABA, and because activity and export of metabolites as a result of ABA application may be involved in an observed physiological response, we believe that the responses of plants to applied ABA should be interpreted with these points in mind.

Based on the results presented here, we conclude that two control mechanisms may be operating at two levels to modify ABA concentrations within the plant. pH may be the major factor for ABA distribution at the subcellular level. The pH differential across membranes regulates the localized ABA concentration changes, such as between compartments within a cell or between adjacent cells. However, over-all ABA levels among different parts of the whole plant seem to be regulated by metabolism. ABA concentrations and effects manifest themselves throughout the integrated plant system by controlling rates of synthesis and rates and direction of metabolism (conjugation vs. catabolism).

Acknowledgment. Isolation and identification of radiolabeled metabolites were carried out in Dr. Mark L. Brenner's lab at the University of Minnesota, St. Paul, with the cooperation of Dr. Mich Hein, presently at Monsanto Company (St. Louis, Missouri).

Literature Cited

1. Tanner, W. Ber Deutsch Bot Bd. 1980, 93, 349-51.
2. Tietz, A. Ludewig, M.; Dingkhun, M.; Dorrfling, K. Planta. 1981, 152, 557-61.
3. Wyse, R. E.; Daie, J; Saftner, R. A. Plant Physiol. 1980, 65, S-120.
4. Daie, J.; Wyse, R. E. Plant Physiol. 1983, 72, 430-33.
5. Daie, J.; Wyse, R. E.; Hein, M.; Brenner, M. Plant Physiol. 1984, (In press).
6. Willenbrink, J. Can. J. Bot. 1980, 58, 816-20.
7. Heilmann, B.; Hartung, W.; Gimmler, H. Z. Pflanzenphysiol. 1980, 97, 67-8.
8. Kaiser, W.M.; Hartung, W. Plant Physiol. 1981, 68, 202-06.
9. Noriaki, K.; Maruta, I; Sugahara, K. Plant Cell Physiol. 1980, 21, 817-28.
10. Milborrow, B.V. In "Phytohormones and Related Compounds--A Comprehensive Treatise."; Letham, D. S.; Goodwin, P. B.; Higgins, T. J. V., Eds. Biomedical: North Holland, 1978; Vol I, p. 295.
11. Astle, M. C.; Rubery, P. H. Planta 1980, 150, 312-20.
12. Walton, D. C. Ann. Rev. Plant Physiol. 1980, 31, 453-543.
13. Zeevaart, J. A. D. Plant Physiol. 1980, 66, 672-78.
14. Daie, J; Campbell, W. F. Plant Physiol 1981, 67, 26-9.
15. Milborrow, B. V. Proc. 10th Inter. Conf. Plant Growth Substances, 1979, p. 262.
16. Zeevaart, J. A. D. Plant Physiol. 1983, 71, 477-81.
17. Jones, H. G. Nature. 1978, 271, 267.
18. Harrison, M.A.; Walton, D. C. Plant Physiol. 1975, 56, 250-54.

RECEIVED February 22, 1984

13

Effects of Bioregulators on Growth and Toxin Formation in Fungi

FREDERICK W. PARRISH

Southern Regional Research Center, Agricultural Research Service, U.S. Department of Agriculture, New Orleans, LA 70179

SUSAN B. JONES

Eastern Regional Research Center, Agricultural Research Service, U.S. Department of Agriculture, Philadelphia, PA 19118

In stationary culture on media containing glucose or sucrose as carbon source, growth of some aflatoxin-producing strains of Aspergillus flavus and A. parasiticus was not affected by PIX bioregulator, but production of aflatoxins was inhibited. Formation of citrinin by a strain of Penicillium citrinin was inhibited by the addition of PIX to yeast extract-sucrose medium or rice grain. No effect of PIX was observed on the formation of secalonic acid D by a strain of Penicillium oxalicum growing on yeast extract-sucrose medium or rice grain. The formation of penicillic acid by a strain of Aspergillus sulfureus growing on the sucrose-based medium was slightly inhibited in the presence of PIX. Transmission electron microscopy showed larger cells with thicker cell walls in the PIX-treated mycelium compared to an untreated control. Differences were also apparent in the lipid reserves and mitochondria.

Bioregulators play an important role in agriculture because of their effects on functions such as root development, flowering, etc. The effects of man-made, agrichemical bioregulators have been concerned mainly with growth, yield, and composition of a wide variety of plant species (1,2). Few studies have been reported on the effects of bioregulators on microorganisms. Erwin and coworkers (3,4) reported that growth retardants, including N,N-dimethyl piperidinium chloride (PIX - BASF), when applied to California cotton, reduced the severity of Verticillium wilt symptoms. However, the main reason for applying PIX to cotton is to decrease growth in the upper region of the plant to produce a more compact conical form. This allows the cotton plants to be spaced closer together giving greater yield. PIX has also been used to shorten the height of winter wheat thereby reducing lodging and increasing the yield (5). Beneficial effects of application of PIX have also been found with citrus, grapes, potatoes, and onions (2).

Our interest in PIX stemmed from studies of this bioregulator on sorghum performed by Dr. Harold Gausman (USDA, Lubbock, TX) and his colleagues at Weslaco, TX. Applications of PIX at a rate of 1.4 oz/acre caused a statistically significant increase in the protein content of sorghum (11.88% to 12.47%), and a concomitant decrease in starch content (74.47% to 73.48%). Amino acid analysis showed no differences between the grains from treated and control plants. This raised the question as to whether addition of PIX to culture media would increase the production of cellulase enzymes by *Trichoderma reesei*. The naturally occurring bioregulators, indole-3-acetic acid and gibberellic acid, at concentrations of 10^{-5} M, double the production of protein and cellulase with Trichoderma species grown on carboxy methyl cellulose (6). Therefore, the effect of PIX on *T. reesei* was investigated.

EFFECT OF PIX ON *TRICHODERMA REESEI*

Commercial PIX solution was sterilized by filtration through a 0.45 micron Millipore filter and added to heat-sterilized, standard Trichoderma medium (7) containing 0.5% carboxy methyl cellulose. The final concentration of PIX was 1 mg/ml (6.68 mM). After inoculation with *Trichoderma reesei* QM 6a spores, the culture media were shaken for 10 days at 28 C. The control medium, with no addition of PIX, produced cellulase enzyme, but no growth of the Trichoderma occurred in the medium containing PIX. When PIX was added to the culture medium 2 days after inoculation, growth was arrested. Incidentally, ear and kernel rot of corn by *Trichoderma reesei* has been reported recently (8), and it would be of interest to examine the effect of PIX on this disease in corn.

FUNGAL TOXINS

A serious problem involving fungi and animal feed was encountered in 1960, when 100,000 young turkeys died in the course of a few months (9). The disease proved to be non-infectious, and was eventually recognized as being caused by aflatoxins produced by *Aspergillus flavus* or *Aspergillus parasiticus* growing on animal feed. The toxins can be produced by these fungi in corn, cottonseed, peanuts, and other commodities, and they affect pigs and calves in addition to poultry. In the years since 1960, the efforts of many scientists have been directed to various aspects of the aflatoxin problem resulting in thousands of published papers. Despite these efforts, serious problems of aflatoxins in feed remain.

At Southern Regional Research Center, studies of fungal toxins comprise a major program. Some of the toxins being studied, and the fungi responsible for their formation, are shown in Table I. The fungi which produce these toxins are plant pathogens and postharvest saprophytes in the class Hyphomycetes. Major emphasis is on the four aflatoxins from *Aspergillus flavus* and *Aspergillus parasiticus*.

TABLE I. Fungal Toxins from some Aspergilli and Penicillia

Toxin	Source
Aflatoxins	Aspergillus flavus Aspergillus parasiticus
Citrinin	Penicillium citrinum Penicillium viridicatum
Penicillic Acid	Penicillium spp. Aspergillus spp.
Secalonic Acid D	Penicillium oxalicum

AFLATOXINS

Culture conditions can influence the extent of aflatoxin production. For example, yields of aflatoxin are greater when Aspergilli are grown on glucose/ ammonium nitrate medium for 7 days at 25°C or 29°C than on Czapek's medium, which is based on the same concentration of glucose (10). With a semisynthetic medium based on 2% yeast extract and 20% sucrose, some Aspergilli which produce aflatoxins in still culture do not do so in shake culture. Liquid media rather than solid media were chosen for most of these experiments because fungal growth is generally slower on solid media, and can lead to decreased levels of mycotoxin production. In addition, isolation of fungal toxins from solid media, e.g., corn and peanuts, often requires additional steps to remove constituents extracted from the agricultural commodity, e.g., oils and fats, which would otherwise interfere with the chromatographic separation and determination of the toxins.

Temperature, moisture content, and surface area of substrates have been shown to be important factors in fungal growth and aflatoxin formation (11,12). The amounts formed on liquid media at 25°C are often 150-200 times greater than the amounts formed at 15°C.

For studies of aflatoxin formation, glucose/ammonium nitrate was selected as the synthetic medium (13) and yeast extract (2%)/sucrose (20%) as the semisynthetic medium (14). Fungi were grown for 6 days at 29°C in shake or static cultures. No adjustment of the initial pH of the medium was made. Experiments were replicated three times.

At the end of the 6-day growth period, a volume of chloroform equal to that of the culture medium was added to each flask. The flask and its contents were heated and magnetically stirred until the chloroform vapor reached the mouth of the flask. The mixture was cooled to room temperature then filtered through a sintered glass funnel. The mycelium was washed with chloroform, and the washings were added to the filtrate. The weight of mycelium was determined by drying the mycelial mats for 24 hr at 70°C.

The filtrate and washings were transferred to a separatory funnel, and after shaking, the chloroform layer was removed. The aqueous layer was re-extracted with an equal volume of chloroform. These extractions were sufficient to remove the aflatoxins from the mycelium and the culture medium. The combined chloroform extracts were evaporated to dryness at 40°C in vacuum (15).

The residue which remained after evaporation of the chloroform was dissolved in a known volume of chloroform and aliquots were applied to a 20 cm x 20 cm thin-layer plate of silica gel together with known amounts of aflatoxin B1, B2, G1, and G2 standards. The plate was developed for 40 min in ether followed by 20 min in ether/methanol/water (96/3/1). The amounts of the individual aflatoxins were determined by measuring the fluorescence with a spectrodensitometer connected to a data handling system. The reproducibility of the assay was ± 6%.

In shake culture on glucose/ammonium nitrate medium, Aspergillus parasiticus NRRL 2999 produced aflatoxins in the control medium, but none in the presence of 50 μg/ml of PIX. When PIX was added at the time the culture media were inoculated with fungus spores (ca. 0.5 x 10^6 spores), no difference in mycelial weight was found between the control and PIX-treated cultures. The same result was observed when addition of PIX was made 24 hr after addition of dormant spores to the culture medium. By this time the dormant spores had undergone transformation to the vegetative mycelial stage. These results show that the effect of PIX is not due simply to prevention of swelling and germination of dormant spores or to an effect on the extent of mycelial growth, as has been observed with so many other inhibitors of aflatoxin formation.

The same effects of PIX were also observed when the concentration of PIX in the culture media was increased to 400 or 4000 μg/ml. These higher levels of PIX were tested because Bennett and coworkers (16) found a concentration dependence of acetone or ethanol on aflatoxin formation. They found in a glucose-containing resting cell medium, which lacks a nitrogen source and allows aflatoxin synthesis, that Aspergillus parasiticus NRRL A-16,462 was markedly inhibited (99%) in aflatoxin formation by the presence of 1 M acetone, but with 0.01 M or 0.1 M acetone, aflatoxin formation was stimulated 2.3-fold. Similar results were observed at the same concentrations of ethanol; 95% inhibition at 1 M, and 1.6-fold stimulation at 0.01 or 0.1 M.

Identical results for the effect of PIX on growth and aflatoxin formation were found with Aspergillus flavus 1000-A when PIX was added at 0, 24, or 48 hr after inoculation of the culture medium. In addition, on yeast extract/ sucrose medium in static culture, aflatoxin formation by Aspergillus parasiticus NRRL 2999 was inhibited by PIX. However, in shake culture, this organism did not produce aflatoxins in the control medium which had no PIX addition.

Similar results of inhibition of aflatoxin formation by PIX were found on glucose/ammonium nitrate medium in shake or static culture with Aspergillus parasiticus SU-42, with two other strains of Aspergillus flavus (SRRC 31 and SRRC 37), and another strain of Aspergillus parasiticus (SRRC 235) in static culture on yeast extract/sucrose medium.

It is noteworthy that, when the PIX-containing medium was heat-sterilized, the inhibitory effect of PIX on aflatoxin formation was not observed.

INHIBITORS OF AFLATOXIN FORMATION

A wide variety of chemicals have been shown to inhibit aflatoxin synthesis on synthetic, semisynthetic or solid media of agricultural commodities. The list includes p-aminobenzoic acid, sulfanilamide, anthranilic acid, potassium sulfite, and potassium fluoride (17); mercaptoethanol (18); caffeine (19); the organophosphate insecticides NALED and DICHLORVOS (20-23); and unidentified compounds in a chloroform extract of carrot roots (24). Many of these inhibitors, e.g., p-aminobenzoic acid, also markedly inhibit fungal growth in a parallel fashion. However, DICHLORVOS, mercaptoethanol, and PIX, which inhibit aflatoxin formation at similar concentrations (Table II), do not strongly inhibit fungal growth.

TABLE II. Comparison of inhibition of aflatoxin formation and mycelial growth by DICHLORVOS, mercaptoethanol, and PIX

Compound	Conc. (µg/ml)	% Inhibition	
		Aflatoxin	Growth
DICHLORVOS	100	99	29
Mercaptoethanol	80	97	30
PIX	50	100	0

A possible explanation of "inhibition" of aflatoxin production which needs to be considered for DICHLORVOS, mercaptoethanol, and PIX, is that in the presence of these compounds the aflatoxins are degraded. Degradation of aflatoxins by mycelia of Aspergilli (25) and by potassium bisulfite (26) has been reported.

If the mechanism of action of PIX is inhibition of aflatoxin biosynthesis, it would be possible to investigate the enzymes and intermediates that are affected by following the approaches described by Bennett and Lee (27). Aflatoxins are produced by a polyketide pathway wherein the carbon skeleton is formed from acetate, with the methoxy methyl group being derived from methionine (28,29). Methods available to study biosynthetic pathways include the use of labelled precursors, blocked mutants, metabolic inhibitors, and cell-free systems (27).

TRANSMISSION ELECTRON MICROSCOPY

Another approach to an understanding of the behavior of fungi with respect to production of metabolites is to examine the fungi by microscopy in order to observe morphological and ultrastructural differences.

Photomicrographs and transmission electron micrographs of *Aspergillus parasiticus* NRRL 2999 grown for 6 days at 27 C on yeast extract/ sucrose medium were obtained. The mycelia were fixed in 2% glutaraldehyde in 0.05 M sodium cacodylate buffer (pH 7.0) for 5 hr at 25°C, and post-stained in 1% osmium tetroxide in the same buffer. Fixed samples were dehydrated through graded aqueous acetone, including en bloc staining in 2% uranyl acetate/ 70% acetone for 1 hr, and embedded in Spurr low-viscosity resin. Thin sections were stained with aqueous uranyl acetate and lead citrate and observed in a Zeiss EM 10 at 60 kV. The transmission electron micrographs are shown in Figures 1,2. Control mycelium from shake culture (Figure 1a), which produces no aflatoxins, showed a densely-staining cytoplasm with many round lipid inclusions. The organelles do not stand out, and no extracellular, fibrillar material is present.

Control mycelium from static culture (Figure 1b), which produces aflatoxins, was different from the control shake mycelium (Figure 1a). The cells appear to be actively metabolizing, and the mitochondria look well-formed. The lipid inclusions were modified so that they have dark material at boundaries, which may be lipid-soluble product. The lipid reserves were not great, and there was a lot of apparent activity in membrane-bound organelles of both degradative and synthetic types.

Mycelium produced in static culture in the presence of sufficient p-aminobenzoic acid to inhibit aflatoxin formation (Figure 2a), was severely affected by p-aminobenzoic acid, and it was difficult to find a live cell to photograph. The ultrastructure was indistinct, the mitochondria were poorly developed, and there were no lipid reserves.

Mycelium from the PIX-treated fungus (Figure 2b), which did not produce aflatoxins, showed cells which were larger than the other 3 samples and the cell walls were thick. There were considerable lipid reserves and some black inclusions, possibly pigments, within vacuoles. There appeared to be a low level of degradative/synthetic activity of the same type as was observed with the control, static mycelium. The mitochondria were also different from the control mitochondria. Those from the PIX treatment have densely-staining matrix and tightly packed cristae (interior laminations) and occur in bizarre shapes. The implications of these observations with respect to active phosphorylation is still under investigation.

CITRININ

Citrinin is a nephrotoxin which can affect swine. Penicillium citrinum 5927 was grown in static culture on yeast extract/sucrose medium or on rice grains with 50% added water. The rice culture medium was acidified to pH 1.5 by addition of 6 M hydrochloric acid, and was extracted twice with chloroform. The chloroform layer was extracted with 0.1 M sodium bicarbonate to remove citrinin, and after acidification the citrinin was extracted into chloroform. This extraction procedure gave a sample suitable for thin layer chromatography. The sodium bicarbonate treatment was not necessary for the chloroform extract of acidified yeast extract/sucrose medium in order to produce a sample for TLC. The assay for citrinin was performed by Stubblefield's procedure (30) for TLC on silica plates containing EDTA. Following development in benzene/acetic acid (95/5), the dried plates were scanned on a spectrodensitometer at an excitation wavelength of 365 nm and emission wavelength of 505 nm. Peak areas were calculated using an automated data system.

Citrinin formation was inhibited by PIX at the only concentration level tested, 3.4 mg/ml in yeast extract/sucrose medium and 17 mg/gram of rice.

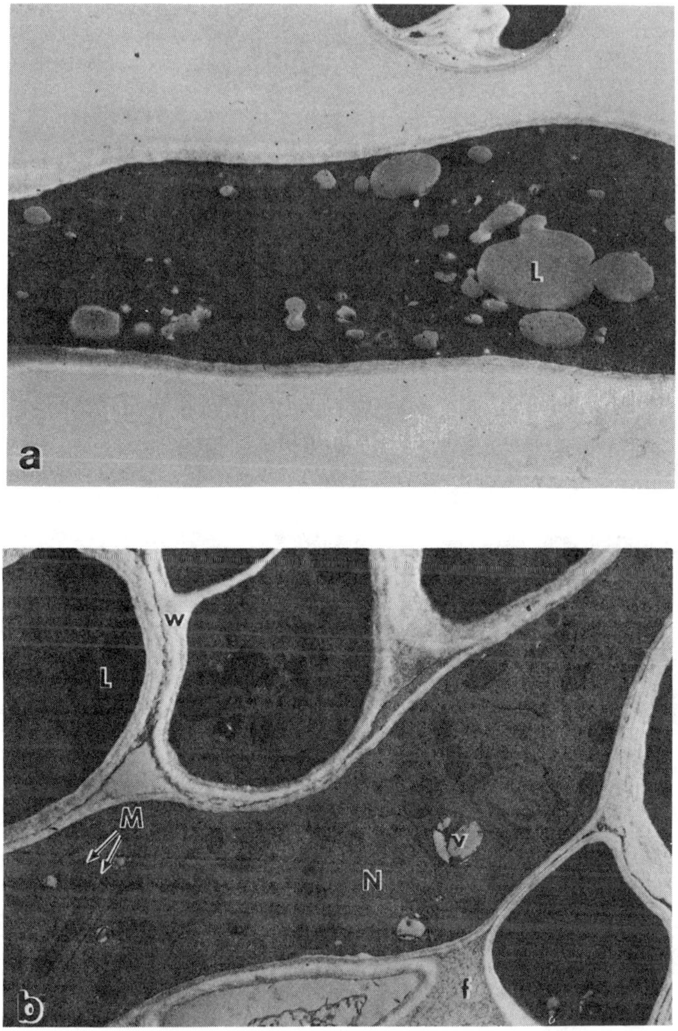

Figure 1. *Aspergillus parasiticus* NRRL 2999 grown in yeast extract/sucrose liquid medium for 6 days at 27 °C. Key: a--shake culture, 135 rpm; no aflatoxins produced; b--still culture, aflatoxins produced, x 10,000.

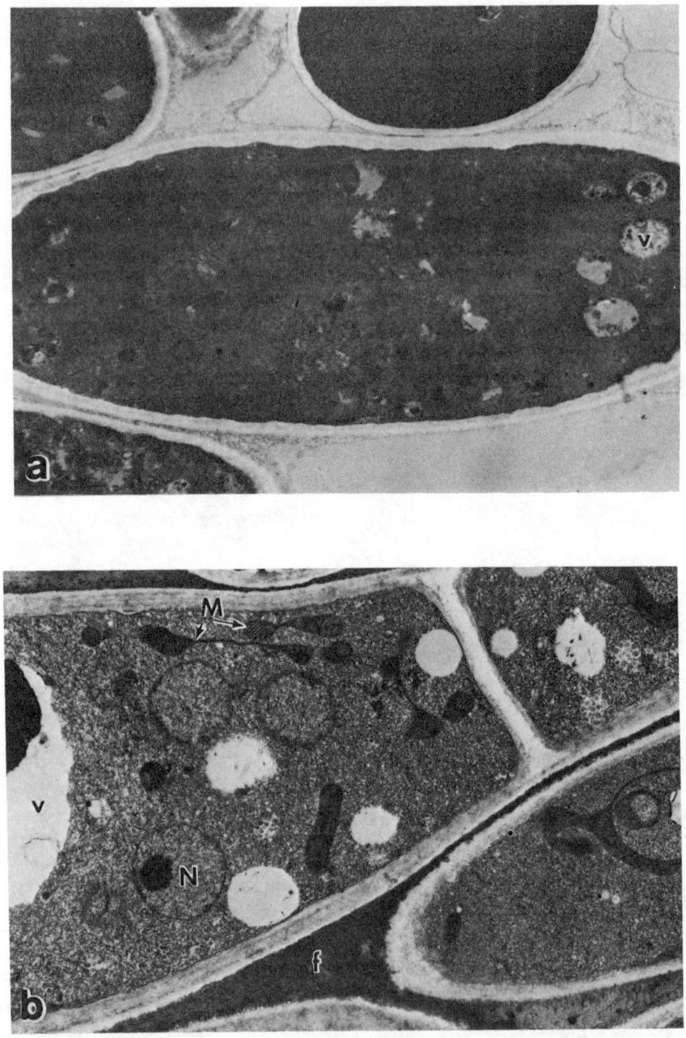

Figure 2. Aspergillus parasiticus NRRL 2999 grown in still cultures with inhibitors of aflatoxin production. Key: a-- p-aminobenzoic acid (PABA), 8 mg/ml, added at 0 time; b-- N,N-dimethylpiperidine chloride (PIX), 50 μg/ml, added at 0 time, x 10,000. Legend: M, mitochnodrion; L, lipid inclusion; N, nucleus; V, vacuole; W, cell wall; f, extracellular fibrous material.

SECALONIC ACID D

A toxin from 5 strains of Penicillium oxalicum was isolated and identified as secalonic acid D by Steyn (31). Extracts of this fungus, which commonly infects corn, are toxic to laboratory and domestic animals.

Penicillium oxalicum 5209 was inoculated into yeast extract/sucrose medium (25 ml) or rice grains (5 g) with 50% added water in static culture at 28°C. The effect of PIX was tested at 3.4 mg/ml in yeast extract/sucrose medium and at 17 mg/gram of rice. Ciegler and coworkers (32) showed that secalonic acid D production with this organism on yeast extract/sucrose medium or on corn was greater in static culture than in shake culture. These authors found levels of secalonic acid D after 14 days inoculation to be about 2 mg/ml on yeast extract/sucrose medium and about 800 mg/kg on rice. Using the same extractant as Ciegler, i.e., methylene chloride, the secalonic acid D was assayed by the high performance liquid chromatographic procedure of Reddy et al (33). This involved removing the methylene chloride by evaporation, dissolving the sample in acetonitrile, and injecting it on to a µBondapak C-18 reverse phase column. The eluant was acetonitrile/water/acetic acid/tetrahydrofuran (4/3/0.5/0.5) in the isocratic mode with detection at 365 nm. Other compounds with similar retention times to secalonic acid D interfered with the assay. However, these interfering compounds were removed readily by extracting the acetonitrile solution, containing secalonic acid D, with hexane.

Secalonic acid D crystallized readily on standing in the acetonitrile solution. Its identity was confirmed by its absorbance characteristics (max 339 nm) as reported by Reddy et al (33), and by mass spectral comparison with authentic material giving the 638 molecular ion and other fragmentation ions.

We found similar levels of secalonic acid D formation on yeast extract/ sucrose medium and on rice to those found by Ciegler et al (32). However, there was no inhibition of secalonic acid D formation by PIX.

PENICILLIC ACID FORMATION

Penicillic acid is a compound produced by a number of Penicillia and Aspergilli which induces fatty liver degeneration in quail and liver cell necrosis in mice (34). Aspergillus sulfureus 97 was grown on yeast extract/sucrose medium for 13 days at 28°C in shake or static culture. Work-up was performed as described by Ciegler et al (34). The acidified (pH 1.5) medium was extracted with chloroform, and the extracts were assayed by thin layer chromatography on silica gel with chloroform/ ethyl acetate/formic acid (60/40/1) as developing solvent. Exposure of the air-dried plate to concentrated ammonia gave a blue fluorescent derivative having Rf 0.5. The effect of PIX addition on penicillic acid formation was to cause 33% inhibition in the presence of 2.4 mg/ml of PIX.

FUTURE WORK

The survey of the effects of PIX on a number of other fungal toxins is continuing with emphasis on the mechanism of PIX action in inhibiting aflatoxin formation. More organisms must be screened on different media, particularly those based on cereals and oilseeds, to determine if the effect of PIX is a general one.

ACKNOWLEDGMENTS

We are grateful to George Cathey (USDA, Stoneville, MS) for supplying commercial PIX; to Timothy Dobson, Beverly Maleeff, and Charlotte Thornton for excellent technical assistance, and to Drs. Ken Ehrlich, Robert Ory, and Falk Rittig for helpful discussions during the course of this work. The mass spectra of secalonic acid D were obtained by Michael Legendre.

LITERATURE CITED

1. York, A. C. Agronomy J. 1983, 75, 663-667.
2. BASF Technical Data Sheet. 1983.
3. Erwin, D. C.; Tsai, S. D.; Khan, R. A. Phytopathology. 1976. 66, 106-110.
4. Erwin, D. C.; Tsai, S. D.; Khan, R. A. Phytopathology. 1979. 69, 283-287.
5. Schott, P. E. Med. Fac. Landbouww. Riyksuniv. Gent. 1979. 44, 853-865.
6. Zamost, B. L.; McLary, D. O. Biotechnology Letters. 1983. 5, 179-184.
7. Mandels, M.; Weber, J. In "Cellulases and Their Applications"; Gould, R. F., Ed.; Advances in Chemistry Series No. 95, American Chemical Society: Washington, D. C., 1969; p. 391.
8. Kumar, V.; Shetty, H. S. Current Sci. 1982. 51, 620-621.
9. Blount, W. P., J. Brit. Turkey Fed. 1961. 9, 52-61.
10. Parrish, F. W.; Wiley, B. J.; Simmons, E. G.; Long, L., Jr. U. S. Army Material Command, Technical Report, Microbiology Series No. 20. 1965.
11. Northolt, M. D.; Bullerman, L. B. J. Food Protection. 1982. 45, 519-526.
12. Delucca, A. J.; Mayne, R. A.; Franz, A. O., Jr.; Ory, R. L. J. Food Protection. 1977. 40, 828-830.
13. Brian, P. W.; Dawkins, A. W.; Grove, J. F.; Hemming, H. G.; Lowe, D.;Norris, G. L. F. J. Exptl. Bot. 1961. 12, 1-12.
14. Davis, N. D.; Diener, U. L; Elridge, D. W. Appl. Microbiol. 1966. 14, 378-380.
15. Hartley, R. D.; Nesbitt, B. F.; O'Kelly, J. Nature. 1963. 198, 1056-1058.
16. Bennett, J. W.; Lee, L. S.; Gaar, G. G. Mycopathologia. 1976. 58, 9-12.
17. Davis, N. D.; Diener, U. L. Appl. Microbiol. 1967. 15, 1517-1518.
18. Gupta, S. K.; Maggon, K. K.; Venkitasubramanian, T. A. Appl. Environ. Microbiol. 1976. 32, 324-326.
19. Buchanan, R. L.; Hoover, D. G; Jones, S. B. Appl. Environ. Microbiol. 1983. 46, 1193-1200.
20. Rao, H. R. G.; Harein, P. K. J. Econ. Entomol. 1972. 65, 988-989.

21. Draughon, F. A.; Ayres, J. C. Appl. Environ. Microbiol. 1981. 41, 972-976.
22. Bennett, J. W.; Lee, L. S.; Cucullu, A. F. Bot. Gaz. 1976. 137, 318-324.
23. Dutton, M. F.; Anderson, M. S. J. Food Protection 1980. 43, 381-384.
24. Batt, C.; Solberg, M.; Ceponis, M. J. Food Sci. 1980. 45, 1210-1213.
25. Doyle, M. P.; Marth, E. H. Mycopathologia. 1978. 63, 145-153.
26. Doyle, M. P.; Marth, E. H. J. Food Protection. 1978. 41, 549-555.
27. Bennett, J. W.; Lee, L. S. J. Food Protection. 1979. 42, 805-809.
28. Biollaz, M.; Buchi, G.; Milne, G. J. Amer. Chem. Soc. 1968. 90, 5017-5019.
29. Biollaz, M.; Buchi, G.; Milne, G. J. Amer. Chem. Soc. 1970. 92, 1035-1043.
30. Stubblefield., R. D. J. Assoc. Offic. Anal. Chem. 1979. 62, 201-202.
31. Steyn, P. S. Tetrahedron. 1970. 26, 51-57.
32. Ciegler, A.; Hayes, A. W.; Vesonder, R. F. Appl. Environ. Microbiol. 1980. 39, 285-287.
33. Reddy, C. S.; Reddy, R. V.; Hayes, A. W. J. Chromatog. 1981. 208, 17-26.
34. Ciegler, A; Mintzlaff, H. J.; Weisleder, D.; Leistner, L. Appl. Microbiol. 1972. 24, 114-119.

RECEIVED April 4, 1984

Biologically Active Natural Products from Fungi: Templates for Tomorrow's Pesticides

HORACE G. CUTLER

Plant Physiology Research Unit, Agricultural Research Service, U.S. Department of Agriculture, Athens, GA 30613

Fungi are a source of unique bioregulators that exhibit a wide range of phytotoxic, or regulatory properties, including fungistatic and bacteriostatic responses. These metabolites offer interesting templates for potential agricultural use. Minor changes in molecules enhance specific activity. A sensitive, relatively simple, primary plant bioassay is used to detect biological activity and novel structures including pergillin, dihydropergillin, chaetoglobosin K, hydroxyterphenyllin and orlandin have been isolated using this assay. Other metabolites have also shown selective activity against higher plants; cytochalasin H controls flowering in tobacco plants; cladosporin diacetate induces chlorosis in corn plants; prehelminthosporol causes necrosis and stunting in corn. Terphenyllin and 6-pentyl-α-pyrone have either fungistatic, or bacteriostatic properties. Biodegradable properties, high specificity and low dosage rates make fungal metabolites, and derivatives, attractive candidates for agricultural use.

Biologically active natural products from fungi, the so-called secondary metabolites, are of primary importance as a source of novel compounds which may be used to control plant growth and development. These compounds tend to be specific in their action on target plants, are generally effective at low concentrations, and tend to be non-persistent in the environment. Furthermore, the compounds offer unique templates for further synthesis of analogs or additional syntheses that start with fermentation products.

Effects of fungally derived natural products on plants include such phenomena as inhibition of vegetative growth, shoot suppression, or inhibition of flowering, depending upon the species treated. Some effects may be classified as truly growth regulatory, others as phytotoxic. Minor changes in an active molecule may increase the specific activity so as to induce chlorosis in select plants or change morphological development. Inhibition of growth is not only limited to higher plants but may also include the fungi and bacteria. The latter, of course, is well documented in the study of pharmaceuticals.

In order to detect biologically active natural products in crude extracts of fungi and to successfully isolate the active principle in pure form, it is necessary to have a highly reliable bioassay. The assay should be broad in response, ideally detecting all manifestations of growth inhibition, or growth promotion (that is, it should clearly exhibit phytotoxicity, fungistatic or bacteriostatic properties, inhibition of cell wall extension and mitosis, and stem curvature). Further, the assay should be relatively simple, highly reliable, quick, and results should be critically reproducible from assay to assay. These characteristics are important because often the chemistry may be apparently anomalous during the isolation of natural products. For example, a major metabolite may appear as a single spot on thin-layer plates when developed and observed under ultra-violet light or treated with a chromogenic agent. But often, the major visible metabolite is not the biologically active material. Instead, the active metabolite may be invisible, or may appear as a very minor spot. The former especially occurs with compounds that have high specific activity; and there are always those metabolites that do not have conjugated double bonds, or suitable chromophores, or are hidden under biologically inactive metabolites even in different developing solvents. Consequently, the primary bioassay is of paramount importance and must be an unerring source of information concerning the whereabouts and nature of the metabolite being tracked during an isolation exercise.

Because of the importance of the bioassay the technique, which has been published elsewhere (1) is nevertheless presented again with some further technical details and discussion. The bioassay is based on an earlier method (2) which has been modified. Wheat seed (Triticum aestivum L. cv Wakeland) are sown on moist sand in plastic trays, covered with aluminum foil, placed in the dark at 22±1°C for 4 days, then the etiolated plants are harvested. Entire shoots are cut from the seed and roots and placed, apex first, into a Van der Weij guillotine. The apical 2 mm are discarded and the next 4 mm are cut and placed in distilled water in a petri dish and soaked prior to use. Generally, compounds to be tested for activity, either in crude extracts or pure, are dissolved in a suitable solvent and a known volume (or concentration) is placed in a test tube and then evaporated to dryness under a stream of nitrogen. When pure compounds are tested they are usually dissolved in acetone up to 7.5 uL acetone/1 mL of final solution. Previous studies have shown that this quantity of acetone is not phytotoxic to either oat first internode segments (3) or wheat coleoptiles (4) Then 2 mL of phosphate-citrate buffer (pH 5.6) containing 2% sucrose (5) are added to the test tube containing the test material and control tubes are prepared with buffer and sucrose only. Next, ten washed coleoptile segments are added to each test tube and the tubes are placed in an inclined roller-tube apparatus that rotates 1/5 rpm. All procedures are carried out under a green safelight at 540 nm, a photomorphogenetically inactive wavelength. After 18-24 hours at 22°C (the time does not have to be exact from test to test because all final measurements are relative to control sections within a specific assay) the sections are removed from the tubes, blotted dry on paper towels, placed on a glass plate, put into a photographic enlarger to project a X3 image and measured (6). Data are subjected to statistical analysis by a multiple comparison procedure (7).

While the mechanics of the bioassay are straightforward there are subtle points that need to be emphasized. First, the sections cut in the guillotine are very accurate. Second, coleoptile segments are hollow cylinders of tissue with cut ends. The surface area that comes into contact with the test solution is larger than it would be with a solid cylinder of tissue, as is the case with oat mesocotyls. Third, because the segments are rotated they are constantly bathed and agitated during the course of the experiment. And fourth, and most important, the success of the bioassay depends on the genotype of the wheat used. This final point cannot be overstressed and a diligent effort to find and choose a genotype for the type of study at hand is essential.

The test of any bioassay as a research tool is directly proportional to the validity of the results obtained. Using the coleoptile bioassay as the primary test, we now discuss some of the novel metabolites that have been isolated from fungi which elicit specific responses in higher plants, or fungi, or bacteria and which may eventually find use as selective herbicides, anti-flowering agents, fungicides and bactericides in agricultural management. While fungal metabolites that exhibit phytotoxic, plant growth inhibitory and other properties have been isolated by other researchers, this discussion is limited to those compounds that we have detected with this bioassay.

Pergillin and dihydropergillin (Figures 1 and 2), were two metabolites isolated from cultures of *Aspergillus ustus* (ATCC 38849) (8,9). This genus and species, though not necessarily this accession, have yielded kojic acid (10) austdiol (11) and austin (12) which are all vertebrate toxins. Crude extracts of the fungus inhibited the growth of wheat coleoptiles; and by a succession of separatory techniques pergillin and dihydropergillin were eventually isolated, though not at the same time. As the authors stated in their isolation of dihydropergillin, ".... since the wheat coleoptile cannot discriminate between individual inhibitors, on a gross scale, it was impossible to determine whether activity in tubes 31-51 was attributable to dilute amounts of pergillin (specific activity data were unavailable at that time) or another inhibitor. It is now obvious that tubes 31-51 contained dihydropergillin. Furthermore, after the crystallization of pergillin from tubes 52-58, there were recoverable traces of dihydropergillin in the supernatant liquid." (9). This statement shows the strength and weakness of the

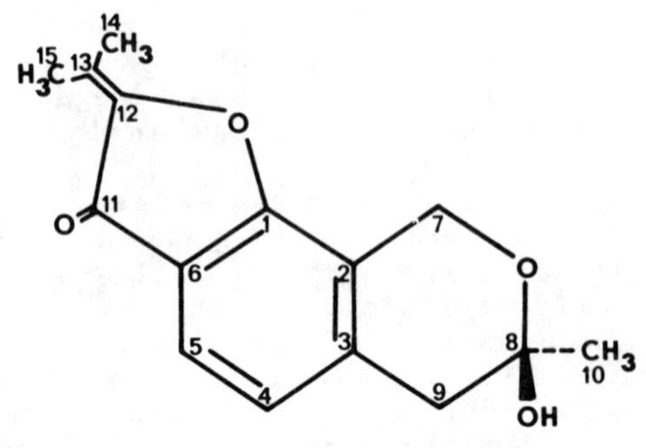

Figure 1 Pergillin

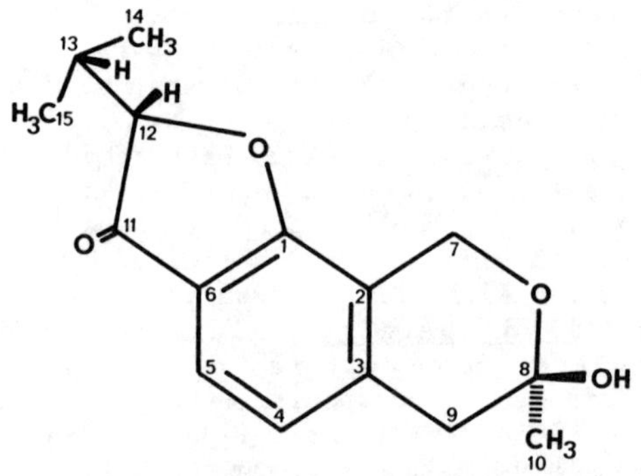

Figure 2 Dihydropergillin

bioassay system, especially when two active metabolites overlap as they exit a column upon chromatography.

Upon examining the molecules of pergillin and dihydropergillin it will be noted that the only difference is that in the latter the C12-13 bond is saturated. This allows the isopropyl group to form a staggered conformation with the furan ring and accordingly alters the biological activity. Thus, etiolated coleoptiles are inhibited 50% at 10^{-3}M with pergillin, while dihydropergillin inhibits 100% (Figures 3 and 4). At 10^{-4}M dihydropergillin significantly inhibits 7% more than pergillin. Minor changes in a molecule may be readily detected by the bioassay and further evidence for this will be presented.

Chaetoglobosin K, was one of the more difficult metabolites to isolate and was obtained from Diplodia macrospora (ATCC 36896) (13) (Figure 5). Initial extractions of the mycelium were made with acetone; and when aliquots of the crude extract were bioassayed, they inhibited coleoptiles some 80% relative to controls. While such a figure is statistically significant, we have generally found that this level of activity does not warrant further isolation effort. As a rule, activity decreases with purity (for example, in constrast to the enrichment process in enzyme purification.) The reason appears to be that crude preparations contain co-factors or related compounds that synergize. However, in this case purification progressed to the point that crystals were obtained with great difficulty. And, exceptionally, the specific activity increased with purity. Chaetoglobosin K, is active at 10^{-7}M in the bioassay (Figure 6). The molecule is unique in that unlike other members of the chaetoglobosin family, there is an additional methyl group at C10 and C11 (14). Subsequent work (15) has disclosed another member of the family with the same substituents which lacks the C6-C7 epoxide, named chaetoglobosin L. It is especially interesting to note that chaetoglobosin K may be considered an indole-3-yl compound and those compounds have historically been promoters of wheat coleoptiles (16). Obviously, chaetoglobosin K is exceptional in this regard. Yahara (17) has tested chaetoglobosin K and other cytochalasins and chaetoglobosins in C3H mouse fibroplastic cells. Chaetoglobosin K, at 20 uM, caused curvature of actin cables, normally present in a straight configuration; but the same concentration did not induce rounding up of fibroplasts. At 0.2 uM membrane ruffling, a function of cell locomotion, was strongly inhibited. Because of very limited quantities, the metabolite could not be tested in other biological systems.

Hydroxyterphenyllin, (2′,5′ -dimethoxy -3,4,3′,4" -tetrahydroxy-p-terphenyl) was a curious metabolite that exhibited biological activity in the primary bioassay (18) (Figure 7). It was obtained from cultures of Aspergillus candidus (ATCC 36008). A different strain of this organism has yielded kojic acid (10). The compound is curious because, on viewing the structure, the first observation is that the metabolite is not a ′typical′ natural product and that it best represents a synthetic product because of the characteristic carbon to carbon bonding between the three rings. Fused ring structures are commonly found natural products and modified biphenyl compounds have been isoalted (vide

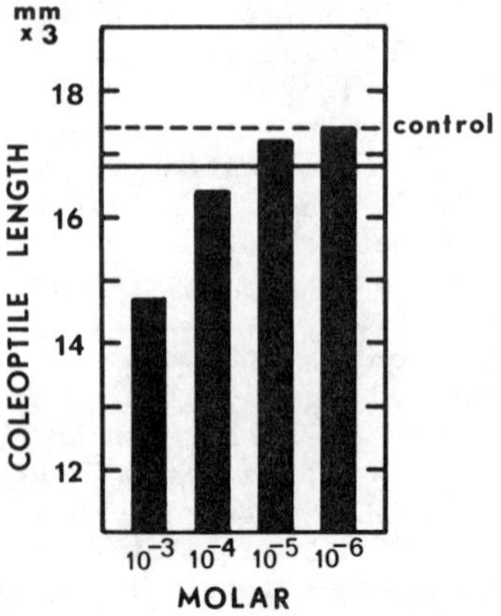

Figure 3 Inhibitory effect of pergillin on the growth of wheat coleoptiles (T. aestivum L., cv. Wakeland). Control: dotted line. Significant inhibition: below solid line (P<0.01).

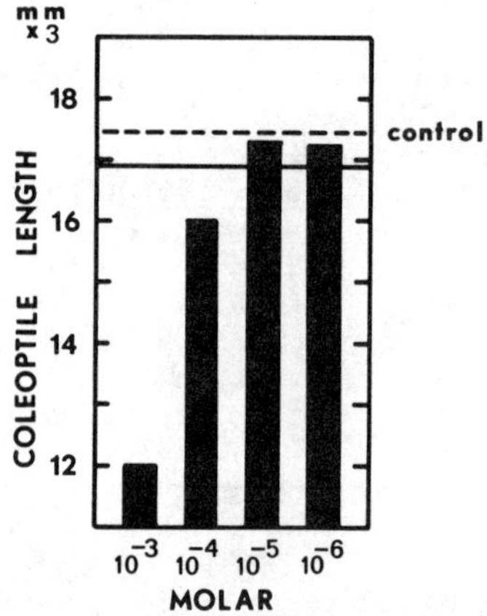

Figure 4 Inhibitory effect of dihydropergillin on the growth of wheat coleoptiles (T. aestivum L., cv. Wakeland). Control: dotted line. Significant inhibition: below solid line (P<0.01).

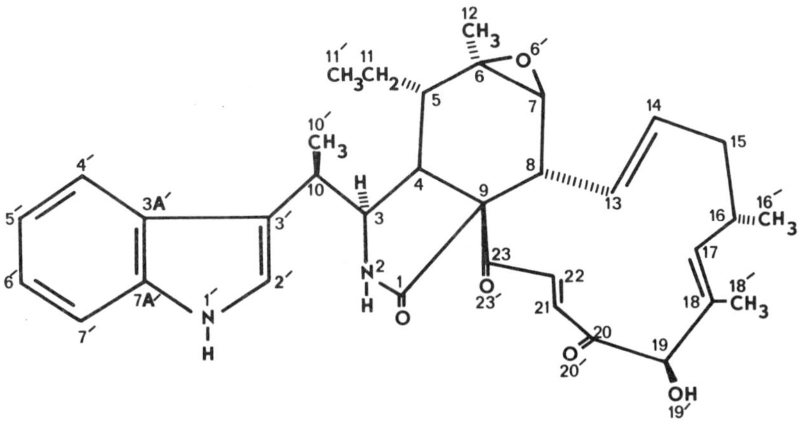

Figure 5 Chaetoglobosin K

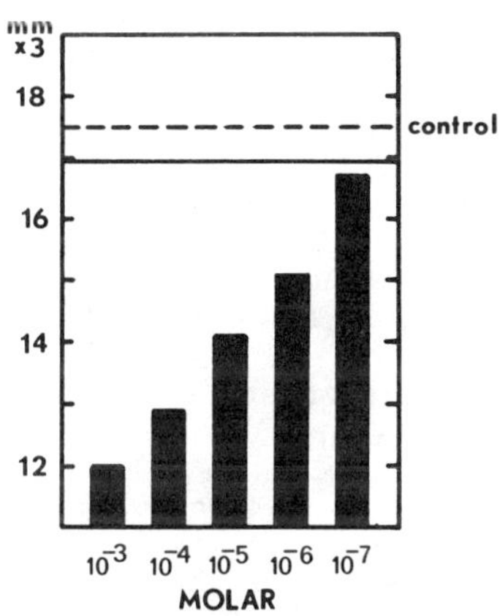

Figure 6 Inhibitory effect of chaetoglobosin K on the growth of wheat coleoptiles (T. aestivum L., cv. Wakeland). Control: dotted line. Significant inhibition: below solid line ($P<0.01$).

infra, orlandin and kotanin for example) but terphenyl structures are exceedingly rare. Paradoxically, our later studies show that substituted terphenyl derivatives are difficult to make and microorganisms are superior in their ability to synthesize these structures. Terphenyllin (I) (19) was isolated in conjunction with hydroxyterphenyllin (II) in our isolation, and both compounds were excellent models for determining relative biological activity. Based on the coleoptile assay, terphenyllin was active only at 10^{-3} M and inhibited 35% relative to controls. The addition of the extra hydroxyl group at the 3 position (hydroxyterphenyllin) dramatically increased the activity so that coeloptiles were inhibited 100, 42 and 8% at 10^{-3}, 10^{-4} and 10^{-5} respectively (Figure 8). By synthesizing terphenyllin tetracetate (III) it was shown that the hydroxyl groups are necessary for biological activity:the tetraacetate was inactive. Questions arise as to whether the addition of extra hydroxyl groups or their positioning increases biological activity or not. Unfortunately, the synthesis is difficult, but it is quite possible that other fungi may produce analogs that can be used as starting materials to produce derivatives. The importance of the molecule lies in the information that terphenyllin was toxic to HeLa cells at 3.2 ug/mL (20). Hence potential chemotherapeutics exist.

Orlandin, a bicoumaryl compound isolated from _Aspergillus niger_ (ATCC 36626), was also of interest because of the biological activity-structural relationship (21). Orlandin (I) closely resembles kotanin (II) (22,23) a metabolite isolated from _A. clavatus_ found contaminating rice, which was implicated in the sudden death of a male youth in Baan Kota, Thailand. The only difference between the two compounds is that in orlandin there are hydroxyl groups at C7, C7′ whereas in kotanin these groups are both methoxy (Figure 9). Consequently, orlandin inhibits the growth of coleoptiles 100, 100 and 35% at 10^{-3}, 10^{-4} and 10^{-5} M respectively, while kotanin is not inhibitory (Figure 10). Conversely, kotanin has an LD_{100} of 62.5 mg/Kg and orlandin has no effect at 125 mg/Kg in day-old chicks.

So far the discussion has centered on new structures that have been discovered using the wheat coleoptile bioassay as a diagnostic tool, and the observations that minor structural changes in molecules change biological properties. We now focus on compounds that exhibit activity in the coleoptile biaossay and also elicit responses in higher plants.

Cytochalasin H was isolated from _Phomopsis_ sp. (4) and _P. paspalli_ (24,25) and was inhibitory in the coleoptile assay at 10^{-3}, 10^{-4}, 10^{-5} and 10^{-6} M (84, 80, 76 and 10% relative to controls) (Figures 11,12). (±) abscisic acid, the standard inhibitor used by us in the assay, is active at 10^{-5} M (18). In addition, cytochalasin H induced curving and slight twisting of coleoptile sections. Experiments were conducted on six-week-old tobacco seedlings (_Nicotiana tabacum_ L.) that had been greenhouse-grown under short daylength and night temperatures of 10°C to induce flowering. Plants were treated at rates of 1.22; 0.12 and 0.01 oz/acre with aqueous sprays of the metabolite. Within 29 days, treated plants were inhibited 87, 81 and 53% at each concentration, respectively. By 43 days, controls were in full bloom and the corollas were

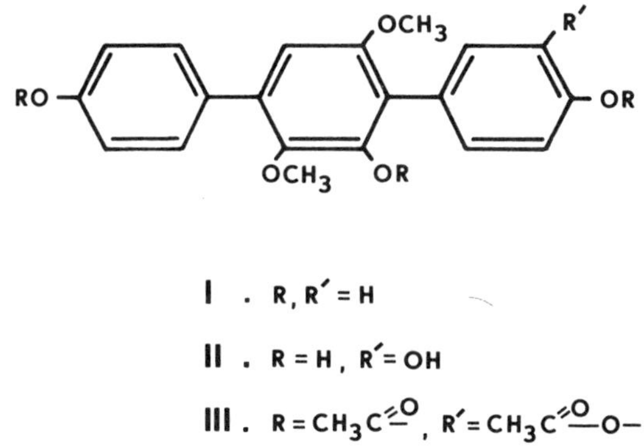

Figure 7 Hydroxyterphenyllin

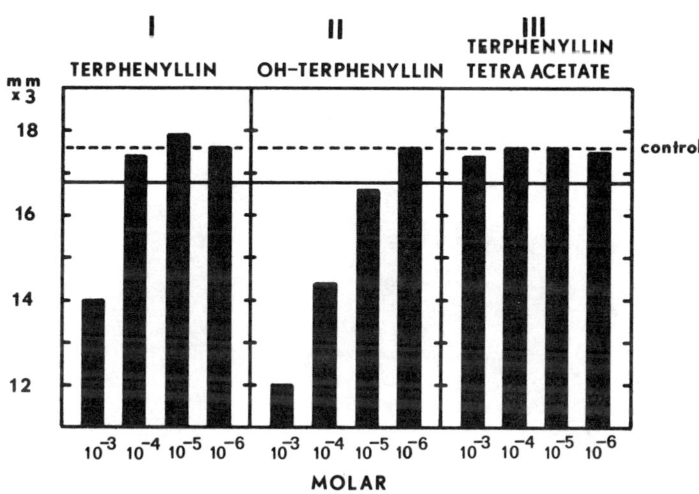

Figure 8 Inhibitory effect of terphenyllin, hydroxyterphenyllin and terphenyllin tetraacetate on the growth of wheat coleoptiles (*T. aestivum* L., cv. Wakeland). Control: dotted line. Significant inhibition: below solid line ($P<0.01$).

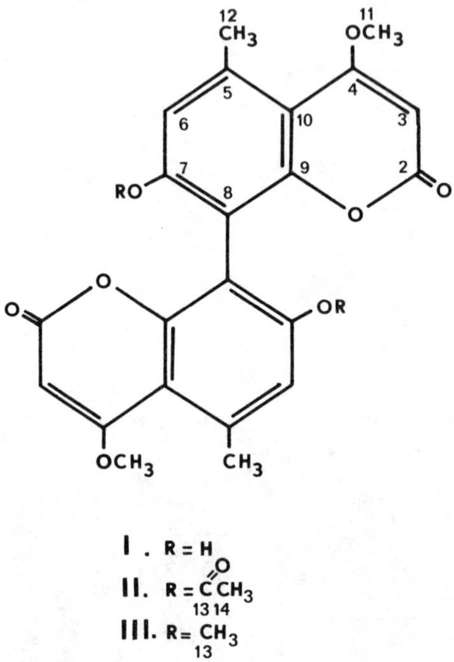

Figure 9 Orlandin

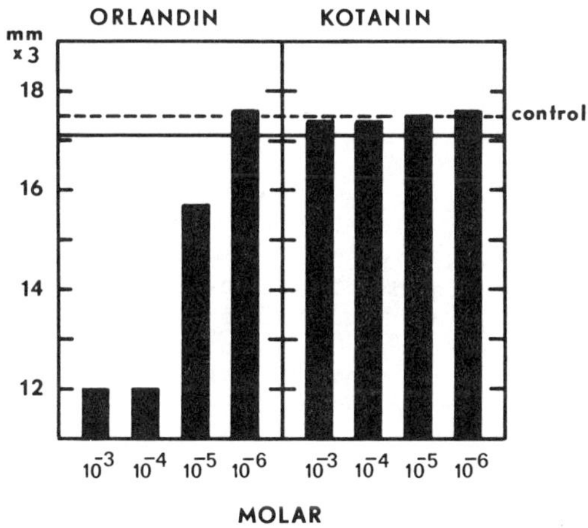

Figure 10 Inhibitory effect of orlandin and kotanin on the growth of wheat coleoptiles (T. aestivum L., cv. Wakeland). Control: dotted line. Significant inhibition: below solid line (P<0.01).

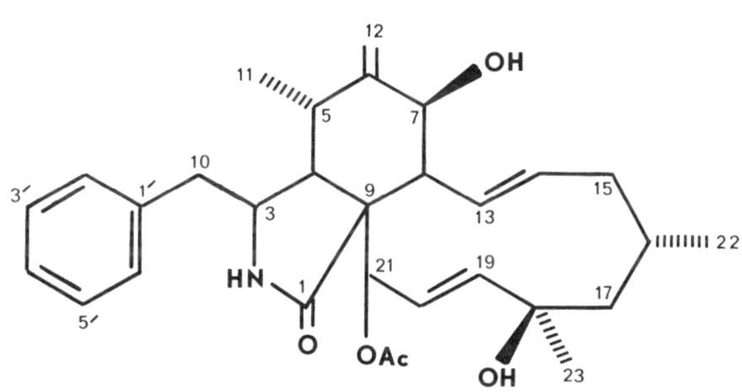

Figure 11 Cytochalasin H

drooping. At 0.01 oz/acre plants were in the early flowering stage and corollas had not opened fully; while the 0.12 oz/acre treated plants were still immature racemes. The 1.2 oz/acre treated plants were still vegetative. Another phenomenon was the appearance of olive colored patches (possible temporary chimeras) on leaves, 4 days after treatment, indicating a possible genetic change in the tissues. Neither corn seedlings (Zea mays L.) nor bean plants (Phaseolus vulgaris L.) were affected by cytochalasin H, with one notable exception. The first true leaves of bean plants rolled longitudinally at all concentrations forming tight cylinders.

Agronomically, early flowering in tobacco fields is a recurrent problem in N. Florida and S. Georgia when warm days (>21°C) are followed by cool nights (2-20°C) in February and March. Correction requires expensive stoop labor and each plant has to be pinched back to one axillary shoot.

Synthesis of the 7-acetoxy; 2,7-diacetoxy; and 2,7,18-triacetoxy derivatives of cytochalasin H reveals something of the functional groups necessary for biological activity (26). The hydroxyl group at C18, and to a lesser extent the NH group at N2, are necessary to obtain plant growth inhibition, while the C7 hydroxyl group is necessary to induce toxic effects in day-old chicks.

Cladosporin was 'rediscovered' using the coleoptile bioassay (27); but, as was stated earlier, the assay does not discriminate between inhibitors or phytotoxins (Figure 13). However, two facts came to light initially. The stereochemistry of cladosporin had not been elucidated, neither had it been previously isolated from Aspergillus repens De Bary (ATCC 38646). During the course of preparing the peracetate for nuclear magnetic resonance spectroscopy enough was prepared for testing on plants. Surprisingly, both cladosporin (I) and cladosporin diacetate (II) were equally active in inhibiting wheat coleoptiles at 10^{-3}, 10^{-4} and 10^{-5} M so that the data from each were superimposable (Figure 14). When six-week-old tobacco plants were treated with cladosporin and cladosporin diacetate there was no response, but treatment of corn plants produced marked leaf chlorosis at 10^{-2} and 10^{-3} M within 24 hours only with the diacetate. Within 48 hours there was chlorosis, necrosis and stem collapse with the 10^{-2}M diacetate treatments. After 12 days, the growth of 10^{-2} and 10^{-3}M treated plants was inhibited. The apparent bleaching of chloroplasts in corn is an interesting topic for further study, and further questions arise. Why did not tobacco chloroplasts behave similarly? What is the nature of the specificity in corn? Is it possible to make pesticides based on the cladosporin structure?

Prehelminthosporol was isolated from the plant pathogen Dreschlera sorokiana (ATCC 42957) (1) which causes destruction of cereal crops in North America (Figure 15). Earlier, it was suggested that helminthosporol was the probable toxin responsible for crop failure, but later it was shown that helminthosporol and helminthosporal are probably artifacts derived from prehelminthosporol and prehelminthosporal (28). In the primary bioassay both prehelminthosporol (I) and the synthetic,

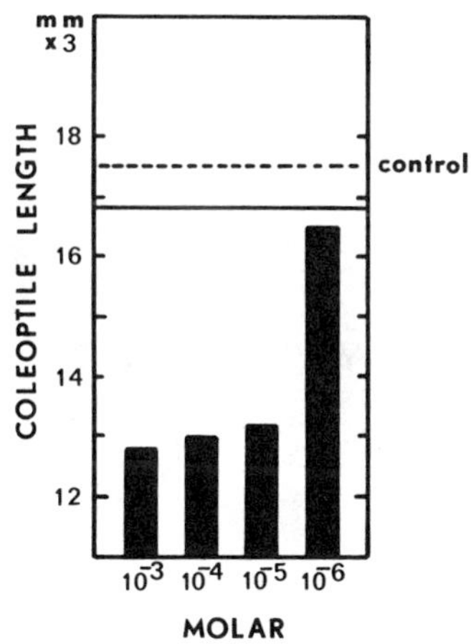

Figure 12 Inhibitory effect of cytochalasin H on the growth of wheat coleoptiles (<u>T. aestivum</u> L. cv. Wakeland). Control: dotted line. Significant inhibition: below solid line (P<0.01).

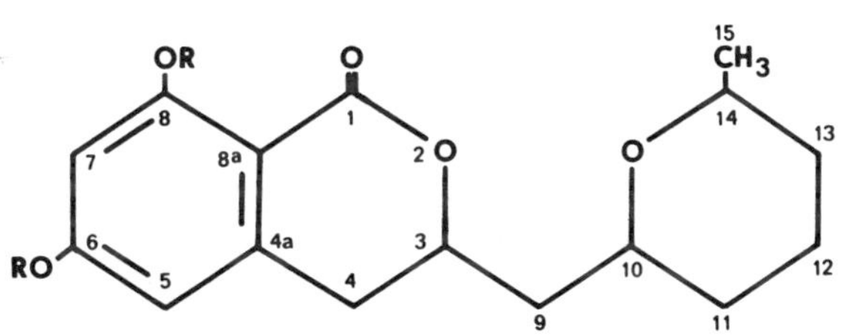

I. R=H
II. R=CH_3CO

Figure 13 Cladosporin

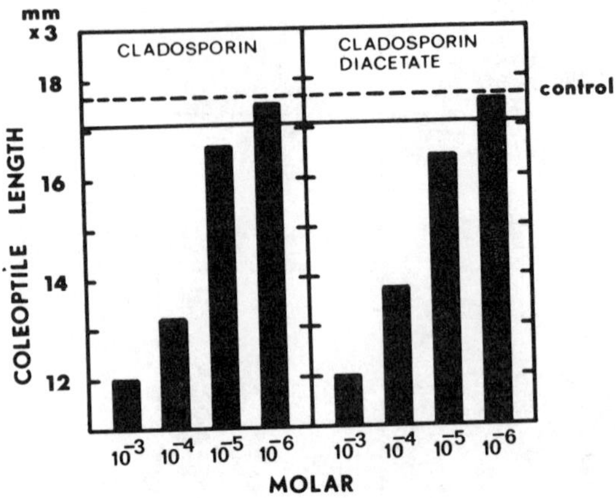

Figure 14 Inhibitory effect of cladosporin and cladosporin diacetate on the growth of wheat coleoptiles (<u>T. aestivum</u> L., cv. Wakeland). Control: dotted line. Significant inhibition: below solid line ($P<0.01$).

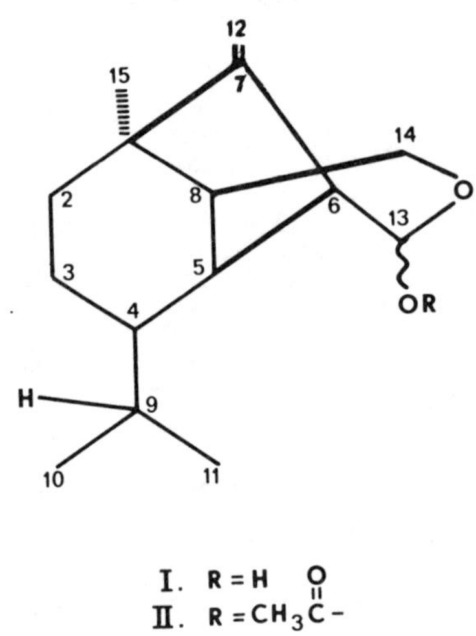

Figure 15 Prehelminthosporol

prehelminthosporol acetate (II), inhibited wheat coleoptiles at 10^{-3} and 10^{-4} M, while at 10^{-5} M there was significant growth promotion which resembled an auxin-like response (Figure 16). Tests in other higher plants induced specific phytotoxicity and plant growth regulator effects. Neither prehelminthosporol nor its acetate induced any response in tobacco plants; but in corn, 10^{-2} M solutions of prehelminthosporol elicited necrosis within 24 hours, and by 48 hours all plants showed some chlorosis and necrosis at 10^{-3} and 10^{-4} M. Prehelminthosporol acetate induced a less severe chlorosis and necrosis at 10^{-2} and 10^{-3}. With bean plants there was an initial response at 24 hours with prehelminthosporol at 10^{-2}M so that approximately 21% of the first true leaves were inverted because of bending of the pulvinus (an auxin-like response). By 48 hours the inversions had increased to 33%. The acetate did not induce bending.

Recently, we have found that the wheat coleoptile bioassay also detects fungistatic, or bacteriostatic compounds. Generally, there is significant inhibition of coleoptile segments at 10^{-3}M and occasionally, but not always, inhibition at 10^{-4} M with these classes of compounds. The end point of activity is reached quite quickly. Further details are the subject of a communication in preparation.

6-Pentyl-α-pyrone was isolated as an aromatic liquid from *Trichoderma viride*. Crude extracts of the organism inhibited coleoptile growth and the inhibition was associated with the presence of a celery odor. The compound was inactive against higher plants (bean, corn and tobacco) but inhibited the growth of mycelia in *Aspergillus flavus* in dilutions of 1:40 in petri dish assays. It was also inhibitory to *E. coli* in dilutions of 1:20. Thus the potential utility of this molecule is based on the following facts. It is a relatively simple molecule and was originally synthesized as a flavoring compound (29) but it has since been found as naturally occurring in peach essence (30) and *T. viride* (31). It is apparently non-toxic, although hard toxicological data are unavailable. Because of these properties, 6-pentyl pyrone, or analogs may be useful in controlling fungi in stored commodities, especially *A. flavus* strains that produce aflatoxin. Another perfuming agent, β-ionone, has been described as an inhibitor of fungi (32).

Further studies with terphenyllin and hydroxyterphenyllin, which were discussed earlier, have shown that both compounds are equally as active as bacteriostatic agents against *Bacillus subtilis*. The peracetate has not yet been tested against either fungi or bacteria.

Any statement concerning the use of natural products as potential agents for use in agriculture, either directly or indirectly as templates for further synthesis, may be met with skepticism and the criticism that such a statement is a cliche´. Only time will vindicate, or condemn this approach to the use of natural products. However, two points should be considered. First, that there are natural products that do exert subtle and not so subtle effects on plant growth and expression. And as a bare minimum, our primary bioassay has helped in the elucidation of several new fungal metabolites. Second, there is an example in

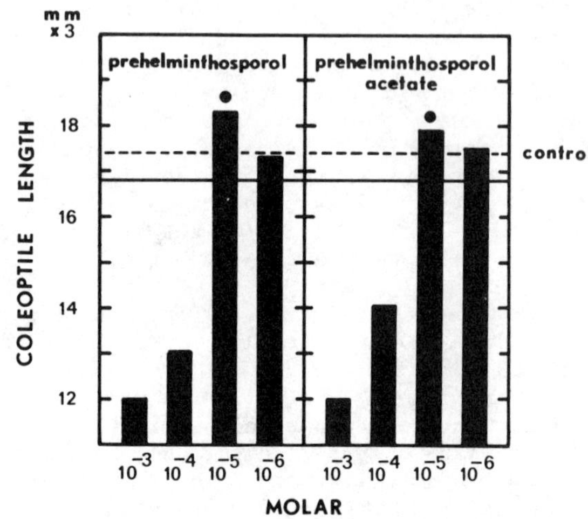

Figure 16 Effects on the growth of wheat coleoptiles (T. aestivum L., cv. Wakeland) by prehelminthosporol and prehelminothosporol acetate. Control: dotted line. Significant inhibition: below solid line ($P<0.01$). Significant promotion: (·) ($P<0.01$).

which a natural product that was shown to have activity in our bioassay, and subsequently in higher plants (33) was synthesized, and homologs were made. Moniliformin (3-hydroxycyclobut-3-ene-1,2-dione), which was initially isolated from Fusarium moniliforme, became the template for cyclobutene-3,4-dione herbicides. Specifically, 1-octyloxycyclobutene-3,4-dione and other derivatives, controlled Setaria italica (L.) Beauv., Lolium perenne L., Sinapsis alba L. and Stellaria media (L.) Villars as postemergence herbicides. Swiss patent 609,836 (Chem. Abstracts 90:198882 p) was granted to this class of compounds based on plant growth regulating and herbicidal activity.

Literature Cited

1. Cutler, H. G.; Crumley, F. G.; Cox, R.H.; Davis, E. E.; Harper, J. L.; Cole, R. J.; Sumner, D. S. J. Agric. Food Chem. 1982, 30, 658.
2. Hancock, C. R.; Barlow, H. W.; Lacey, H. J. J. Exp. Bot. 1964, 15, 166.
3. Cutler, H. G. Plant Cell Physiol. 1968, 9, 593.
4. Wells, J. M.; Cutler, H. G., Cole, R. J. Can. J. Microbiol. 1976, 22, 1137.
5. Nitsch, J. P.; Nitsch, C. Plant Physiol. 1956, 31, 94.
6. Cutler, H. G.; Vlitos, A.J. Physiol. Plant. 1962, 15, 27.
7. Kurtz, T. E.; Link, R. F.; Tukey, J. W.; Wallace, D. L. Technometrics 1965, 7, 95.
8. Cutler, H. G., Crumley, F. G.; Springer, J. P.; Cox, R.H.; Cole, R. J.; Dorner, J. W.; Thean, J. E. J. Agric. Food Chem. 1980, 28, 989.
9. Cutler, H. G.; Crumley, F. G.; Springer, J. P.; Cox, R. H. J. Agric. Food Chem. 1981, 29, 981.
10. Cole, R. J.; Cox, R.H. In "Handbook of Toxic Fungal Metabolites"; Academic Press: New York, 1981. p 759.
11. Steyn, P. S. Tetrahedron 1973, 29, 107.
12. Chexal, K. K.; Springer, J. P.; Clardy, J.; Cole, R. J.; Kirksey, J. W.; Dorner, J. W.; Cutler, H. G.; Strawter, B. J. J. Am. Chem. Soc. 1976, 98, 6748.
13. Cutler, H. G.; Crumley, F. G.; Cox, R. H.; Cole, R. J.; Dorner, J. W.; Springer, J. P.; Latterell, F. M.; Thean, J. E.; Rossi, A. E. J. Agric. Food Chem. 1980, 28, 139.
14. Springer, J. P.; Cox, R. H.; Cutler, H. G.; Crumley, F. G. Tetrahedron Lett. 1980, 21, 1905.
15. Probst, A.; Tamm, C. Helv. Chimic. Acta 1982, 65, 1543.
16. Phillips, I.D.J. In "Introduction to the Biochemistry and Physiology of Plant Growth Hormones", McGraw-Hill: New York, 1971; McGraw-Hill Series in Organismic Biology, p. 11.
17. Yahara, I.; Harada, F.; Sekita, S.; Yoshihira, K.; Natori, S. J. Cell Biol. 1982, 92, 69.
18. Cutler, H. G.; Lefilco, J. H.; Crumley, F. G.; Cox, R.H. J. Agric. Food Chem. 1978, 26, 632.
19. Marchelli, R.; Vining, L. C. J. Antibiot. 1975, 28, 328.
20. Takahashi, C.; Yoshihira, K.; Natori, S., Umeda, M. Chem. Pharm. Bull. 1976, 24, 613.

21. Cutler, H. G.; Crumley, F. G.; Cox, R. H.; Hernandez, O.; Cole, R. J.; Dorner, J. W. J. Agric. Food Chem. 1979, 27, 592.
22. Büchi, G.; Klaubert, D. H.; Shank, R. C.; Weinreb, S. M.; Wogan, G. N. J. Org. Chem. 1971, 36, 1143.
23. Büchi, G.; Luk, K. C.; Kobbe, B.; Townsend, J. M. J. Org. Chem. 1977, 42, 244.
24. Patwardhan, S. A.; Pandey, R. C.; Dev, S.; Pendse, G. S. Phytochem. 1974, 13, 1985.
25. Pendse, G. S. Experientia 1974, 30, 107.
26. Cox, R.H.; Cutler, H. G.; Hurd, R. E.; Cole, R. J. J. Agric. Food Chem. 1983, 31, 405.
27. Springer, J. P.; Cutler, H. G.; Crumley, F. G.; Cox, R.H.; Davis, E. E.; Thean, J. E. J. Agric. Food Chem. 1981, 29, 853.
28. De Mayo, P.; Williams, R. E.; Spencer, E. Y. Can. J. Chem. 1965, 43, 1357.
29. Nobuhara, A. Agr. Biol. Chem. 1969, 33, 1264.
30. Sevenants, M. R.; Jennings, W. G. J. Food Sci. 1971, 36, 536.
31. Collins, R. P.; Halim, A. F. J. Agric. Food Chem. 1972, 20, 437.
32. Wilson, D. M.; Gueldner, R. C.; McKinney, J. K.; Lievsay, R. H.; Evans, B. D.; Hill, R. A. J. Am. Oil Chem. Soc. 1981, 58, 959A.
33. Cole, R. J.; Kirksey, J. W.; Cutler, H. G.; Doupnik, B. L.; Peckham, J. C. Science 1973, 179, 1324.

RECEIVED February 22, 1984

15

Effects of 1,1-Dimethylpiperidinium Chloride on the Pests and Allelochemicals of Cotton and Pecan

P. A. HEDIN, J. N. JENKINS, J. C. McCARTY, JR., J. E. MULROONEY, and W. L. PARROTT

U.S. Department of Agriculture, Mississippi State, MS 39762

A. BORAZJANI and C. H. GRAVES, JR.

Department of Plant Pathology and Weed Science, Mississippi State University, Mississippi State, MS 39762

T. H. FILER

Southern Forest Experiment Station, Stoneville, MS 38776

The growth regulator, PIX (mepiquat chloride = 1,1-dimethyl-piperdinium chloride), when applied to cotton (Gossypium hirsutum L.) and pecan (Carya illinoensis Koch), caused internode shortening. PIX did not elicit an increase in resistance in cotton to the tobacco budworm [Heliothis virescens (Fab.)], or in pecan to pecan scab [Cladosporium caryigenum (Ell. et Lang) Gottwald]. Also, changes in content of four known allelochemicals (condensed tannins, gossypol, anthocyanins, flavonoids) for these pests were minimal. An unexpected finding was the increase in content of several nutritional factors that may be related to greater, rather than lesser, growth of tobbaco budworm larvae feeding on cotton tissues.

Plant growth regulators have an important role in the growth and developmental processes of plants. In cotton, (Gossypium hirsutum L.) termination of late season fruiting has been achieved with potassium 3,4-dichloroisothiazole-5-carboxylate, thus depriving the pink bollworm [Pectinophora gossypiella (Saunders)] of food and oviposition sites (1). Of perhaps greater importance would be the control of insects during the growing season. Plant growth regulators have been shown to increase the biosynthesis of certain secondary plant constituents that, in turn, decrease plant attack by insects. Gibberellic acid for example, elicits increased terpene biosynthesis in citrus (Citrus sp.), thus decreasing attack by fruit flies (Anastorpha sp.) (2,3).

Phytoalexin is a general term for compounds that are induced, whether by an infectious agent or by a chemical compound such as a bioregulator (4). More phytoalexins are synthesized when the plants are subjected to stress.

Plant growth regulators have been both isolated from and applied to cotton. Recently, 1,1-dimethyl-piperidinium chloride (mepiquat chloride = PIX) has been found to control undesirable vegetative growth, and to promote boll set (5,6). There have been recent reports about the effects of PIX on insect pests of

0097–6156/84/0257–0171$06.50/0

cotton. Zummo et al. (7) reported less plant damage, decreased bollworm [Heliothis zea (Boddie)] growth, and 10-20% increased terpenoids, tannins, and astringency (biological tannin) in a Texas field plot test. Ganyard (8) in North Carolina, observed a 23% decrease in bollworm damage in PIX treated cotton.

There do not appear to be any reports about the application of PIX to pecans (Carya illinoensis Koch). Many other plant growth regulators have been applied to a number of fruit and nut trees and their effects recorded. Borazjani (9) has reviewed much of this literature. Increased fruit set and chlorophyll, and decreased internode lengths, have often been observed.

This study was initiated to determine whether PIX and other growth regulators enhanced a) the intrinsic resistance of cotton to insects by increasing allelochemics, and b) the resistance of pecan to pecan scab. These crops and their pests were selected for study because they were conveniently available, and are the subjects of comprehensive on-going breeding programs at this location.

Materials and Methods 1/

1982 PIX Cotton Test. Pix was applied to Stoneville 213 (ST 213) cotton on July 6 and July 21 at the rates of 0, 0.4, 1.2 and 3.6 liter/hect A.I. There were six replications. Plant tissues were harvested immediately before the first and second treatments, and also at 2 weeks following the second treatment. They were freeze-dehydrated and ground, then held at -20C in cap-closed vials until analyzed. Second instar tobacco budworm larvae housed individually in plastic cages in the laboratory were fed plant tissues (either leaves or buds) for three days. Rate of growth, in mg/day, was determined.

Analysis of Allelochemicals. Analyses for gossypol and related terpenoid aldehydes were performed on cyclohexane/ethyl acetate/acetic acid: 500/500/1 (CHEA) extracts of plant tissue by the phloroglucinol reaction (2% in abs. ETOH/Con HCl: 1/1, stand 1 hr) with subsequent spectrometic analysis at 550 nm. The concentration was determined by comparison with data obtained from authentic gossypol, and is expressed as gossypol equivalents. Condensed tannin analyses were performed on 70% aqueous methanol (MW) extracts of tissue. The anthocyanidin chromophore was developed by boiling 1 hr with n-butanol/HCl:95/5. The concentration was determined by comparison with the color obtained at 550 nm, from a purified cotton condensed tannin sample, the structure of which has recently been elucidated by Collum et al. (10). The anthocyanin content was determined by measuring the absorbancy at 540 nm of an extract of freeze dehydrated tissue, extracted with methanol/water/HCl: 79/19/3, using the molar extinction coefficient (E) of Cyanidin-3-β-glucoside (11). Flavonoids were determined after extraction of freeze-dehydrated tissue with 70% aqueous acetone. Diphenylboric acid-ethanolamine complex (Natural

1/ Mention of a trademark, proprietary product or vendor does not constitute a guarantee or warranty of the product by the U.S. Department of Agriculture and does not imply its approval to the exclusion of other products or vendors that may also be suitable.

Product reagent A, Aldrich Chem. Co. 1%) in methanol was added, and the chromophore absorptivity at 440 nm was determined and compared to that obtained from a purified sample of isoquercitrin, the most prevalent flavonoid in cotton. AOAC methods were used for the following analyses: Solids (moisture); 14.083, crude fat; 14.019, crude fiber; 14.118, ash; 14.114, and nitrogen free extract (NFE) by difference from 100%.

1981 Pecan Test of Growth Regulators on Pecans. Fifteen mature Van Deman pecan trees located on the Black Belt Experiment Station in Brooksville, MS were used to evaluate the effects of four growth regulators on juglone synthesis and/or accumulation and growth habits. A randomized complete block experimental design with 3 replications was employed, using one-tree plots. Growth regulators for this study were Indole-3-acetic acid (Sigma Chemical Co., St. Louis, MO), Gibberellic acid (GA_3) (Sigma Chemical Co., St. Louis, MO), Para-coumaric acid (Sigma Chemical Co., St. Louis, MO), and PIX (1,1-dimethyl-piperidinium chloride, BASF Co., Limburgerhof, West Germany). A single spray application was made on April 24, 1981, using a hydraulic sprayer (Tricone^R nozzle 10-14 kg/cm^2). Each material was applied to the foliage at a concentration of 100 ppm to the point of "coverage without run-off" (approximately 56.8 L/tree). Sampling procedures for juglone analyses were carried out at three week intervals beginning in May and ending in late September. Two samples of leaves and nuts were taken from individual terminals in widely spaced random locations on each tree. Each sample of leaves and nuts consisted of no less than 10 g. Samples were transported in an ice chest to the laboratory and placed in the freezer.

1982 Test of PIX on Pecans. Twenty-one mature Van Deman pecan trees, located at the Black Belt Experiment Station in Brooksville, MS were used for this study. A randomized complete block design was employed, using one-tree plots and 3 replications. PIX was applied to the foliage by using a hydraulic sprayer (Tricone nozzle 10-14 kg/cm^2) at concentrations of 100 and 200 ppm (ai) to the point of "coverage without run-off" (56.8 L/tree). The first applications were made when first leaves were one-third grown and applications of some treatments were repeated at 3-week intervals. There were seven treatments, including 1, 2, 3 and 4 applications at 100 ppm, 1 and 2 applications at 200 ppm, and a check.

Analyses and Measurement of Pecan, 1981-2. Juglone analyses were accomplished using a technique developed by Hedin et al. (12). Extractions were made from 10 g of leaves or nuts by grinding in chloroform/methanol (2:1). The filtrate from 3 successive extractions of tissue was combined and concentrated to 25 mL by evaporation under vacuum at 50 C. A 1 mL aliquot was banded on a Silica Gel G TLC plate and chromatographed with methylene chloride/pentane (1:3). Juglone was observed as a yellow-orange band at R_f 0.40, scraped into a test tube, eluted from the silica gel with methylene chloride, filtered, and diluted to 10 mL for spectrometric analysis at 420 nm. For comparison, a standard curve was prepared with dilutions of 0.05-0.50 mg/10 mL of an authentic sample of juglone (Aldrich Chemical Co., Milwaukee, WI.). MS analysis showed that the fragmentation patterns of the commercial sample and the sample from pecan leaves were identical. Analyses for condensed tannins were performed as described earlier. Samples were oven dried, ground, and analyzed for P, K, Ca, Mg, Mn, Fe, Zn, Cu, and B using spark emission spectroscopy procedures described by Jones and Warner (13). Total nitrogen was measured by the micro-Kjeldahl procedures of Bremner (14).

Ten terminals were taken from each tree at widely spaced random locations on June 18 and again on September 24, 1981, for the growth habit study. Terminals were wrapped in moist paper and transferred to the laboratory for measurements. Leaf area was measured by means of an Automatic Area Meter (Hayashi Denko Co., Tokyo, Japan). Terminal length, nut weight, nut number, terminal leaf area, specific leaf weight, and number of leaflets/leaf were also measured.

Results and Discussion.

1982 PIX Cotton Test.

In evidence that PIX had elicited its expected agronomic effects, the internode lenths of ST-213 cotton were decreased by approximately 25%; similar reductions were reported by Namken and Gausman (5). When cotton plant tissues were harvested and fed to tobacco budworm larvae in the laboratory for 3 days, the growth rate was significantly increased by 55% at the level of 3.6 l/hect; .59 mg/mg larva/day vs control; .38 mg/mg larva/day (15).

The concentrations of allelochemicals in terminals and buds are listed in Tables I, II, III, and IV and show changes by the end of the experiment. The allelochemicals measured (tannins, gossypol, anthocyanin, and flavonoids) have been reported to contribute to resistance in cotton to the tobacco budworm (16). Gossypol levels in buds, but not in the leaves, increased significantly 4 weeks after treatment of the plants with the 3.6 l/hect. rate. The average concentration of gossypol, over time, was also significantly increased by this treatment. Leaf tannin and flavonoids were significantly decreased 4 weeks after applying PIX. Except for a small decrease in leaves and buds at the highest level of treatment, there was no significant effect of PIX on the anthocyanin level.

The increase in larval growth rate may be partly explained by increases in cotton terminal nutrients (Table V), perhaps coupled with the decrease of flavonoids in leaves. Increases were obtained for minerals (26%), protein (14%), and lipids (42%). Decreases were obtained for nitrogen free extract (9%), and more notably for crude fiber (16%).

In summary, increases were recorded in PIX treated cotton leaves for larval growth rate, protein, lipid, minerals, and gossypol in buds; there were decreases in leaf flavonoids and tannins, crude fiber, and internode distance (Table VI), but no effect on anthocyanins. The growing year, 1982, had more rainfall, than average, so that the cotton was not drouth stressed. The higher rate of larval growth may have been partially attributable to the higher nutrient concentrations and, perhaps, to the lower flavonoid level in PIX treated plants. The decreased larval growth rates observed in Texas by Zummo et al. (7) and in North Carolina by Ganyard (8) (1982) were obtained during adverse growing seasons when the cotton was stressed. Additionally, they grew larvae on growing plants and we used excised terminal leaves.

Table I. Effect of PIX on tannin levels of leaves and buds of ST 213 cotton over time, test commenced July 6

Treatment	Prior to treatment		2 weeks later [1]		4 weeks later [2]		$\bar{x}$ over time	
	leaf	buds	leaf	buds	leaf	buds	leaf	buds
ℓ/hect.	%							
0	8.21a [3]	8.62a	13.78a	9.64a	21.93a	10.67a	14.64a	9.64a
0.4	7.86a	8.06a	13.55a	9.73a	19.14ab	9.49a	13.52a	9.09a
1.2	10.59a	8.07a	14.79a	8.63a	16.34 bc	10.19a	13.91a	8.96a
3.6	8.84a	10.35a	14.40a	9.74a	13.68 c	11.33a	12.32a	14.47a

1/ Sampled immediately before second PIX application.

2/ Four weeks after first application, two weeks after second PIX application.

3/ Means within a column followed by the same letter are not significantly different at 0.05 level, as determined by Duncan's Multiple Range Test.

Table II. Effect of PIX on gossypol levels of terminal leaves and buds of ST 213 cotton over time; test commenced July 6

Treatment	Prior to treatment		2 weeks later[1]		4 weeks later[2]		$\bar{x}$ over time	
	leaf	buds	leaf	buds	leaf	buds	leaf	buds
ℓ/hect.	- - -	- - -	- - -	- - - %	- - -	- - -	- - -	- - -
0	0.43a[3]	0.35 b	0.29a	0.14a	0.24a	0.22a	0.32a	0.24a
0.4	0.40a	0.34ab	0.28a	0.16a	0.29a	0.25a	0.32a	0.25a
1.2	0.42a	0.28ab	0.32a	0.21a	0.24a	0.31a	0.33a	0.26a
3.6	0.48a	0.24a	0.36a	0.22a	0.29a	0.34 b	0.37 b	0.26a

1/ Sampled immediately before to second PIX application.

2/ Four weeks after first PIX application, two weeks after second PIX application.

3/ Means within a column followed by the same letter are not significantly different at the 0.05 level, as determined by Duncan's Multiple Range Test.

Table III. Effect of PIX on anthocyanin levels of terminal leaves and buds of ST 213 cotton over time; test commenced July 6

Treatment	Prior to treatment		2 weeks later[1]		4 weeks later[2]		$\bar{x}$ over time	
	leaf	buds	leaf	buds	leaf	buds	leaf	buds
ℓ/hect.	- - - - - - - - - -		- - - - - - - - - -		% - - - - - - - - - -		- - - - - - - - - -	
0	1.15a[3]	0.10a	0.53a	0.07a	0.53a	0.59a	0.74a	0.25a
0.4	0.79a	0.09a	0.40a	0.11a	0.50a	0.56a	0.56a	0.25a
1.2	0.94a	0.09a	0.43a	0.08a	0.45a	0.52a	0.61a	0.23a
3.6	0.76a	0.10a	0.50a	0.09a	0.43a	0.44 b	0.56 b	0.21a

1/ Sampled immediately before to second PIX application.

2/ Four weeks after first PIX application, two weeks after second PIX application.

3/ Means within a column followed by the same letter are not significantly different at the 0.05 level, as determined by Duncan's Multiple Range Test.

Table IV. Effect of PIX on flavonoid levels of terminal leaves and buds of ST-213 cotton over time, test commenced July 6

Treatment	Prior to treatment		2 weeks later[1]		4 weeks later[2]		$\bar{x}$ over time	
	leaf	buds	leaf	buds	leaf	buds	leaf	buds
ℓ/hect.	%							
0	2.76a[3]	0.84a	2.93a	1.27a	3.09a	1.35a	2.92a	1.15a
0.4	2.71a	0.79a	2.75a	0.13a	2.57 b	1.12a	2.67 b	1.00a
1.2	2.74a	0.77a	2.96a	1.02a	2.51 b	1.46a	2.73 b	1.08a
3.6	2.67a	0.85a	2.67a	1.13a	2.13 c	1.39a	2.50 b	1.12a
LSD .05	ns	ns	ns	ns	0.37		0.22	ns

1/ Sampled immediately before to second PIX application.

2/ Four weeks after first PIX application, two weeks after second PIX application.

3/ Means within a column followed by the same letter are not significantly different at the 0.05 level, as determined by Duncan's Multiple Range Test.

Table V. Effects of PIX on ST-213 proximate composition of cotton terminals after 28 days

Treatment	Ash	Crude Protein	Crude Fat	Crude Fiber	NFE
l/hect.			%		
0	5.95	21.41	3.16	7.63	47.47
1.2	7.16	22.50	3.92	6.61	46.50
3.6	7.52	24.38	4.48	6.39	43.07

Table VI. Summary; Effects of PIX (3.6 l/hect.) on ST-213 cotton terminal content after 28 days

Increases		Decreases		No Change
Protein	14%	Flavonoids	31%	Gossypol
Lipid	42%	Tannins	38%	Anthocyanins
Minerals (ash)	26%	Internode Distance	25%	
Larval Growth Rate	55%	Crude Fiber	16%	

1981 Pecan test. The effects of the growth regulators on growth habits at 8 and 21 weeks after application are given in Tables VII and VIII. Only the specific leaf weight of differed significantly among the five treatments at 8 weeks. Leaves of trees treated with PIX had a higher specific weight (g/cm^2) than those treated with other chemicals on both observation dates. At the second sampling date, the number of leaflets/leaf for GA_3 and IAA treated leaves was significantly higher than for the other treatments (Table VII). When the data from the two observation dates were combined and analyzed collectively, all sprayed trees showed a significant decrease in terminal length, except those treated with GA. Also, PIX treated trees appeared to have a higher nut set, nut weight, and specific leaf weight for terminal leaves, but only the specific leaf weight of terminal leaves was significantly different at the P=0.05 level.

The increased specific leaf weight with PIX agrees with the findings of Gausman et al. (6) that this growth retardant increases leaf weight and reduces the leaf area due to changes in the mesophyll structure of leaves. Reduced height of cotton was also reported for PIX by (Gausman et al. (6). Since GA_3 and IAA promote new growth, the increase in number of leaflets/leaf seems consistent. The increased length of terminals observed here is consistent with the known elongation of internodes by GA_3 and their ability to cause stem elongation and even to reverse genetic dwarfism in some plants (17).

Juglone in the leaves of PIX and IAA treated trees was significantly higher than in the check leaves 3 weeks following application (Table IX). The juglone level was significantly lower in all treatments at 9 weeks. Although some differences were observed among treatments in the following weeks, none was significant. Juglone levels in nuts for each sampling date are given in Table X. No significant differences were observed among nut tissues until 12 weeks following application, when PIX and IAA-treated

Table VII. Pecan Growth Habit Measurements (6/18/81)

Treatment[1]	Terminal Length	Nut/ Treatment	Nut Wt.	Leaflet/ Leaf	Leaf Wt./ Terminal	Terminal Leaf Area	Terminal Leaf	Terminal Leaf Rachis Length
	cm		g		g	cm^2	mg/cm^2	cm
PIX	7.8 a[2]	2.6 a	1.7	9.8	25.5	298	17 a	34.5
IAA	6.8 a	2.7 a	1.6	9.9	24.6	295	15 b	34.2
GA_2	8.0 a	2.6 a	1.6	9.9	23.8	305	15 b	34.7
P-Coumaric acid	7.6 a	2.4 a	1.4	9.2	25.1	301	14 b	34.0
Check	7.6 a	2.6 a	1.6	9.7	27.1	330	14 b	35.4

1/ 100 PPM to dip off on 4/24 (56.8 L/tree).

2/ Values within a column followed by the same letter are not significantly different, as determined by Duncan's Multiple Range Test, $p=0.05$.

Table VIII. Pecan Growth Habit Measurements (9/24/81)

Treatment[1]	Terminal Length	Nut/ Treatment	Nut Wt.	Leaflet/ Leaf	Leaf Wt./ Terminal	Terminal Leaf Area	Terminal Leaf	Terminal Leaf Rachis Length
	cm		g		g	cm^2	mg/cm^2	cm
PIX	5.6 a[2]	2.3 a	9.8	9.5	25.2	297	16 a	35.3
IAA	7.6 a	1.9 a	8.1	10.1	25.9	308	13 b	34.5
GA_3	9.5 a	1.9 a	8.6	10.3	28.8	336	13 b	38.1
PCA	6.8 a	1.8 a	8.1	9.4	24.0	316	13 b	33.1
check	8.6 a	1.6 a	9.3	9.3	26.4	288	13 b	32.8

1/ 100 PPM to drip off on 4/24 (56.8 L/tree).

2/ Values within a column followed by the same letter are not significantly different, as determined by Duncan's Multiple Range Test, p=0.05.

Table IX. Pecan Leaf Juglone Concentration (mg/g fresh tissue)

Treatment[1]	1981 Sampling Dates						
	May 15	June 4	June 25	July 16	Aug 6	Aug 27	Sept 17
	mg/g fresh weight						
PIX	1.23 a[2]	0.93	0.52	0.37	0.56	0.33	0.25
IAA	1.21 ab	0.76	0.44	0.54	0.58	0.27	0.20
GA_3	0.89	0.85	0.49	0.40	0.60	0.41	0.23
PCA	0.81 bc	0.86	0.70 b	0.62	0.38	0.32	0.34
Check	0.65 c	0.82	1.06 a	0.71	0.55	0.30	0.16

1/ 100 PPM to drip-off on 4/24 (56.8 L/tree).

2/ Values within a column followed by the same letters are not significantly different, as determined by Duncan's Multiple Range Test, p=0.05.

Table X. Pecan Nut Juglone Concentration (mg/g fresh tissue)

Treatments[1]	1981 Sampling Dates					
	June 4	June 25	July 16	Aug 6	Aug 27	Sept 17
PIX	0.10 a[2]	0.16	0.45 ab	0.25 b	0.33 a	0.32 a
IAA	0.12 a	0.12	0.50 a	0.27 b	0.39 a	0.18 b
GA_3	0.13 a	0.12	0.30 bc	0.33 b	0.18 a	0.26 a
PCA	0.11 a	0.17	0.30 bc	0.52 a	0.08 a	0.25 a
Check	0.13 a	0.22	0.30 c	0.30 b	0.21 a	0.29 a

1/ 100 PPM to drip off on 4/24 (56.8 L/tree).

2/ Values within a column followed by the same letters are not sigificantly different, as determined by Duncan's Multiple Range Test, p=0.05.

trees showed significantly higher juglone levels than the controls(Table X). The p-coumaric acid treatment showed a significantly higher juglone content than the other plant growth regulators at 15 weeks following application. Also, on the last sampling date (21 weeks after application), IAA treated nuts showed lower juglone content than the others.

The lack of continued influence of bioregulators on juglone levels may result from dilution by increased plant biomass with time after treatment. IAA moves rapidly from the young green tissue to older tissue, and it is constantly being destroyed by indole-3-acetic acid oxidase (17). Gibberellin is also reported to lose its biological activity gradually after treatment (18); the coumarins are known to bind to sugar to form their glucosides (19). IAA, GA, and p-coumaric acid are reported to stimulate respiration (20), possibly by activating the oxidative enzyme system, thereby enhancing synthesis of phenolic compounds such as juglone. There is no report about the effect of PIX on phenolic pathways.

Delayed juglone accumulation in nut tissues (until 12 weeks after application) suggests that time may be related to translocation of excess material observed earlier in the leaves. Conversely, compounds may have accumulated that were subsequently metabolized amd permitted the increase in juglone.

1982 Pecan Test. Terminal shoot growth of all treated trees was significantly reduced by the first measurement on June 29 (Table XI). However, there were no significant differences from the unsprayed check for any of the other growth measurements that were made on this date.

Table XI. Effect of PIX on growth of mataure Van Deman trees

Treatment Number	Mepiquat Chloride Concentration	Number of Applications	Terminal Shoot Length
			-cm-
1	200 ppm	1	6.93 b[1]
2	200 ppm	2	7.15 b
3	100 ppm	1	6.60 b
4	100 ppm	2	7.13 b
5	100 ppm	3	6.86 b
6	100 ppm	4	6.70 b
7	Unsprayed check	0	9.50 a

Note: Measurements were made on 6/29/82; each value is the average of 3 replications.

1/ ($p = 0.05$) according to Duncan's Multiple Range Test.

Significant differences were recorded for all growth parameters measured in September (Tables XII, XIII). Terminal shoot growth was reduced from 30 to 50% among the various treatments (Table XII). This dramatic effect is significant, since the concentration of PIX used in this experiment was only a fraction of that commonly used on cotton. There were essentially no differences between treated and non-treated trees for number of nuts/terminal or in weight. Otherwise, there was a slight reduction in the number of leaflets/leaf (Table XII) and total leaf

Table XII. Effect of PIX on mature Van Deman pecan trees, as measured on Sept. 22, 1982

Treatment Number	PIX concentration	Number application	Terminal shoot length	Number of nuts/terminal	Nut weight	Number of leaflets/leaf
			(cm)		(g)	
1	200 ppm	1	5.60 b[1]	2.10ab	11.75ab	7.28 b
2	200 ppm	2	6.13 b	2.10ab	11.14ab	7.27 b
3	100 ppm	1	5.95 b	2.36a	12.54a	8.03 b
4	100 ppm	2	5.31 b	1.84 b	12.50a	6.96 b
5	100 ppm	3	4.58 b	2.00ab	10.75 b	7.32 b
6	100 ppm	4	6.30 b	1.93 b	12.32ab	7.64 b
7	Unsprayed check		9.15a	1.92 b	11.82ab	9.36a

Note: Each figure represents an average of 3 replications.

1/ Means within a column not followed by the same letter differ significantly ($p = 0.05$) according to Duncan's Multiple Range Test.

Table XIII. Effect of PIX on leaf characteristics of mature Van Deman pecan trees, as measured on Sept. 22, 1982

Treatment Number	PIX Concentration	Number Applications	Total leaf wt/ Terminal	Average Leaf Area	Specific Leaf Wt.	Average Rachis Length
			(cm)		(gm/cm 2/)	(cm)
1	200 ppm	1	10.28 b [1]	216 d	.013 abc	28.50 bc
2	200 ppm	2	15.43 b	264 b	.016 a	26.26 d
3	100 ppm	1	14.25 b	288 b	.014 abc	28.56 bc
4	100 ppm	2	12.03 b	239 c	.013 abc	29.02 bc
5	100 ppm	3	11.48 b	256 c	.012 c	28.06 c
6	100 ppm	4	14.75 b	278 b	.013 abc	29.78 b
7	Unsprayed check		22.05 a	328 a	.015 ab	32.00 a

Note: Each figure represents an average of 3 replications.

1/ Means within columns not followed by the same letter differ significantly ($p = 0.05$) according to Duncan's Multiple Range Test.

weight/terminal (Table XIII). Reductions in average leaf weight/unit area and average rachis length (Table XIII) were generally more pronounced.

These data suggest that PIX may be useful in reducing pecan tree size. This material may hold potential for those using precocius (early bearing) varieties in close spacing. These compounds delay crowding in these plantings, thus permitting longer periods of high yield before thinning to reduce the crowding effects.

Significant effects on seasonal juglone patterns in both leaves and nuts were noted for all treatments. However, the patterns among the treatments tended to be somewhat erratic and difficult to interpret. The patterns for the two highest rates, (four treatments at 100 ppm and two treatments at 200 ppm) and the check are given in Figure 1. The juglone level of leaves for the unsprayed check is consistent with that found for the Van Deman cultivar in previous studies. The juglone levels for nuts at the two highest rates and the check are given in Figure 2. The greatest effect on juglone levels in nuts may occur near the end of the season.

PIX is systemic and is not readily metabolized in cotton plants (17). Therefore, repeated applications during the growing season might be expected to increase the juglone level, but this trend was not consistent in this study. This may be due to a dilution of PIX by increased plant biomass, or photodegradation of PIX on leaf surfaces. The juglone content was studied because juglone has been shown to be a factor of resistance in pecan to pecan scab. In these PIX treatments, however, the incidence of pecan scab was not significantly reduced.

Tannin levels in leaves increased temporarily for all treatments through 3 and 6 weeks (April 28 and May 18 sampling dates respectively) after first applications (Table XIV). Therefore, following this, no significant differences among treatments were observed, with few exceptions, for the remainder of the season. PIX had no appreciable influence on the foliar levels of the nutrients (N,P,K,Ca,Mg,Mn,Fe,Zn,Cu,and B).

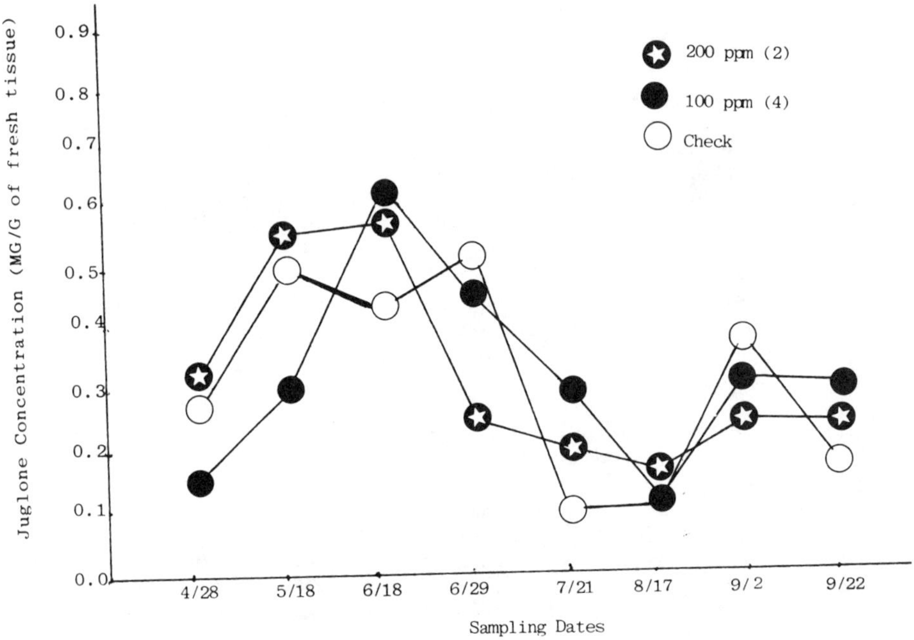

Figure 1. Seasonal Juglone concentrations in leaves of Van Deman pecan trees treated with PIX (200 ppm, 2 applications and 100 ppm, 4 applications) as compared with unsprayed check. Each datum point represents an average of 3 replications.

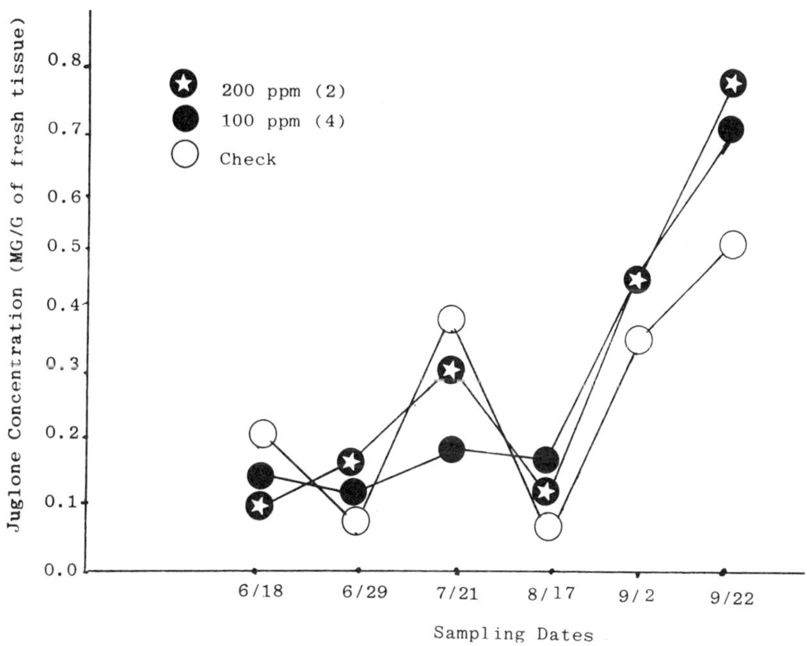

Figure 2. Seasonal juglone concentrations in nuts of Van Deman pecan trees treated with PIX (200 ppm, 2 applications and 100 ppm, 4 applications) as compared with unsprayed check. Each datum point represents an average of 3 replications.

Table XIV. Tannin concentration in leaves of Van Deman pecan trees treated with PIX

Treatment Number	PIX Conc.	# of appl.	Sampling Dates: Apr. 28	May 18	June 8	June 29	July 21	Aug. 12	Sept. 12	Sept. 22
			mg/g fresh wt.							
1	200 ppm	(1)	2.45a [1]	3.01 b	3.57ab	5.04a	4.97ab	5.46 cd	7.07ab	5.39 b
2	200 ppm	(2)	2.52a	2.94 c	3.57ab	4.41ab	3.65 c	5.81 cd	6.44ab	5.32 b
3	100 ppm	(1)	2.27a	2.76 e	3.57ab	4.48ab	3.85 bc	6.44a	6.09 b	7.21a
4	100 ppm	(2)	2.45a	3.01 b	3.36ab	4.87ab	4.20abc	6.02abc	6.58ab	6.02 b
5	100 ppm	(3)	2.73a	3.17a	4.04a	4.20ab	5.37a	6.3 ab	7.28a	5.25 b
6	100 ppm	(4)	2.34a	2.91a	3.20ab	3.99 b	3.64 c	5.25 d	6.58ab	5.25 b
7	Unsprayed check		1.78 b	2.56	3.01 b	4.06 b	4.55abc	5.95abc	6.37ab	6.16 b

Note: Each figure represents an average of 1982 data from 3 replications.

1/ Means within columns not followed by the same letter differ significantly at ($p = 0.05$) according to Duncan's Multiple Range Test.

Literature Cited

1. Kittock, D. L., Arle, H. F., Bariola, L. A. Proc. Beltwide Cotton Prod. Res. Conf., 1975, Jan. 6-8, New Orleans, LA, p. 71.
2. Greany, P. I. Agricultural Research, U.S. Dept. of Agriculture, Washington, DC. 1978. 27, 3.
3. Coggins, C. W., Scora, R. W., Lewis, L. N., Knapp, J. C. F. J. Agr. Food Chem., 1969, 17, 807.
4. Keen, N. T., Bruegger, B. In "Host Plant Resistance to Pests"; Hedin, P. A., Ed., American Chemical Society, Washington, DC, 1977, ACS Symp. Ser. No. 62, pp. 1-26.
5. Namken, L. N., Gausman, H. W. Agricultural Research, U. S. Dept. of Agriculture, Washington, DC, 1978, 27, 11.
6. Gausman, H. W., Rittig, F. R., Namken, L. N., Rodriguez, R. R., Escobar, D. E., Garza, M. V. Proc. Fifth Ann. Mtg., Plant Growth Regulator Working Group, Blacksburg, VA, 1978, June 25-29, p. 137.
7. Zummo, G. R., Benedict, J. H., Segers, J. C. Proc. Beltwide Cotton Prod. Res. Conf., 1983, Jan. 3-6, San Antonio, TX, p. 73.
8. Ganyard, M. C. Successful Farming, May 1982, p. 17.
9. Borazjani, A. Ph.D. Thesis, Miss. State, MS, 1981, 68 pp.
10. Collum, D. H., White, W. H., Parrott, W. L., Jenkins, J. N., Hedin, P. A., Crimley, E. B. Abstracts of Papers, 182nd National Meeting of the Am. Chem. Soc., 1981, New York, NY., PEST 54.
11. Hedin, P. A., Minyard, J. P., Thompson, A. C., Struck, R. E., Frye, J. Phytochemistry, 1967, 6, 1165.
12. Hedin, P. A., Langhans, V. E., Graves, Jr., C. H. J. Agr. Food Chem., 1977, 27, 92.
13. Jones, J. B., Warner, N. H. in Dev. of Applied Spectroscopy Vol. 7A. Plenum Press, N.Y. 1969. p. 152.
14. Bremner, J. M. 1965. Total nitrogen. In C., A. Black, (ed). Methods of soil analysis Part 2. Chemical microbiological properties, Adv. in Agron. 1965, 9, 1149.
15. Mulrooney, J. W. Unpublished data.
16. Hedin, P. A., Jenkins, J. N., Collum, D. H., White, W. H., Parrott, W. L., Mac Gown, M. W. Experientia, 1983, 39, 799.
17. Gausman, H. W., Walter, H., Rittig, F. R., Scobar, D. E., Rodriguez, R. R. Proc. 7th Ann. Mtg., Plant Growth Regulator Group. 1980, Dallas, TX, July 13-16, p. 1.
18. Hedden, P., MacMillan, J., Phinney, B. O. Ann. Rev. Plant Physiol. 1978, 29, 149.
19. Mandava, N. B. "Plant Growth Substances", USDA, Washington, DC, 1979, 310pp.
20. Berrie, A. M. M., Parker, W., Knights, B. A., Hendric, M. R. Phystochemistry, 1968, 7, 567.

RECEIVED April 4, 1984

16

Bioregulation of Host Plant Resistance to Insects

B. C. CAMPBELL, B. G. CHAN, L. L. CREASY,[1] D. L. DREYER, L. B. RABIN, and A. C. WAISS, JR.[2]

Western Regional Research Center, Agricultural Research Service, U.S. Department of Agriculture, Berkeley, CA 94710

Natural and synthetic plant regulators have been used extensively in agriculture. This paper reviews the literature on the effect of these compounds, directly or indirectly, on herbivorous insects and spider mites. Several attempts to use plant bioregulators (PBR) to modify insect resistance in plants have been reported. Experiments in our laboratories indicate that populations of aphids on sorghum and alfalfa, and the growth of Heliothis zea on tomato, corn, and cotton may be reduced with applications of appropriate PBRs. The mechanism of the induced resistance in these crop plants are being investigated.

For the last several years our research group in Albany has been involved in elucidating the chemical interactions between crop plants and their respective insect pests (1-10). While advances in the understanding of the chemical nature of host plant resistance may facilitate plant selection for insect resistance, classical plant breeding remains a slow and laborious task. The modification of plant resistance using plant bioregulators (PBR) to manipulate the biosynthesis, metabolism, and translocation of natural resistance factors and essential insect nutrients in the host should should be a worthy complement to the present methods of pest control. Similarly, the use of plant growth regulators (PGR) to alter the physical and morphological characteristic of the host, such as increasing physical barriers and early termination of vegetative parts, which deny the insect of food and shelter may also be effective in reducing pest populations. A PBR, as defined by Maier and Yokoyama (11), modifies the plant through specific biochemical processes without imparting gross morphological changes that is expected from a PGR, though a particular regulator may have both biochemical and growth modifying activities.

[1]On sabbatical leave from the Department of Pomology, Cornell University, Ithaca, NY 14853
[2]To whom correspondence should be addressed

Many plant regulators, herbicides, and fungicides are presently being used experimentally and commercially in the field. The impact of these compounds, whether beneficial or detrimental, on plant diseases and insect pests should be fully investigated so that their use in an overall strategy of integrated pest management can be optimally exploited.

This publication summarizes and reviews 1. some of the unexpected effects plant regulators have on the population of non-target insects; 2. toxicities of plant regulators to arthropods; 3. reported literature on the bioregulation for host plant resistance to arthropods; and 4. recent progress on induced resistance to insects in our laboratories.

Inadvertent Side Effects of Plant Regulators and Herbicides on Insects

Most of the reported unexpected side effects of plant regulators to insects involved mainly the application of the herbicide 2,4-D (2,4-dichlorophenoxy acetic acid, Ia) on the hosts. As summarized in Table I, the application of 2,4-D may inadvertently alter the populations of herbivorous insects. However, these changes were mostly a result of the effects of 2,4-D on natural enemies of these insects. For example, the increase in number of sugarcane borer *Diatres saccharalis* (Fabricius) on sugar cane was attributed by Ingram and co-workers (12) to the detrimental effect of 2,4-D on the parasitic wasp, *Trichogramma minutum* Riley. Similarly, populations of a number of graminivorous aphids increased in oat fields treated with the herbicide. This was caused, according to Adams and Drew (13), by a reduction in the population of *Hippodamia tredecimpunctata tibialis* (Say), a predatory ladybird beetle of aphids. Retarded germination and reduced vigor of wheat by early application of 2,4-D was considered the factor contributing to increased plant damage by the prairie grain wireworm, *Ctenicera aeripennis destructor* (Brown) (14). The direct toxicity of the herbicide was reportedly responsible for larval mortality of the wheat stem sawfly, *Cephus cinctus* Norton (15). Inadvertent toxicity of herbicides to bees has also been documented (16,17).

Table I. Inadvertent Effects of the Herbicide 2,4-D on Insects.

Plant Regulator	Host Plant	Insect	Effect on Pest Population	Ref.
2,4-D	Sugar cane	Sugarcane borer	Increased	12
2,4-D	Oat	Grain aphids	Increased	13
2,4-D	Wheat	Wireworm	Increased	14
2,4-D	Wheat	Wheat stem sawfly	Reduced	15

Direct Effects of Plant Regulators on Arthropods

Plant regulators may directly effect the growth, development, and fecundity of insects or spider mites. These effects are especially evident when these compounds are applied at dosages higher than normally used on plants. Using a castor leaf bioassay, El-Ibrashy and Mansour (18) demonstrated that chlormequat chloride (II) reduced larval growth and pupal weight of black cutworm, *Agrotis ipsilon* (Hufnagel). Antigrowth activity was also observed when the compound was injected into the larvae. Antifeeding activity of chlorphosphonium chloride (III) to the cotton leafworm, *Prodenia litura* Fabricius was reported by Tahori and coworkers (19) after application of this compound on cotton or wax bean foliages. Synthetic diets containing 0.5 to 2% of either maleic hydrazide (IV) or chlormequat chloride were reported to increase mortality and reduce fecundity of the pea aphid, *Acyrthosiphon pisum* (Harris) (20).

In 1960, Eichmeier and Guyer (21) observed a significant reduction in the population of two spotted spider mite, *Tetranychus urticae*, Koch, grown on snap bean plants that were treated with 15 ppm of gibberellic acid (V). This observation was confirmed later by Rodriguez and Campbell (22). However, scientists at the Anti-Locust Research Center in England demonstrated that gibberellic acid is a necessary dietary constituent for normal maturation in desert locust, *Schistocerca gregaria* (Forskål) (23). They further showed that artificial diets containing chlormequat chloride, a synthetic inhibitor of gibberellic acid biosynthesis, retarded and prevented sexual maturation of both locust and the cotton stainer, *Dystercus cardinallis* Gerth. No significant effect was observed with abscisic acid (VI), a natural antagonist of gibberellic acid.

Recently, Visscher found that both gibberellic acid and abscisic acid added to a grass diet, significantly reduced the fecundity, egg viability and, thus the rate of reproduction of the grasshopper, *Aulocara elliotti* (Thomas) (24). Because these experiments were carried out with diets containing fresh plant material and the plant hormones were applied at unusually high concentrations as compared to the natural endogenous levels of these hormones, it is difficult to rule out the possibility that these observed effects are indirectly due to alteration in the physiological conditions of the grass in the diet and not direct effects on the insect themselves.

We have recently conducted bioassays of several plant regulators against larvae of corn earworm, *Heliothis zea* (Boddie) by incorporating the plant regulators into artificial diets (Table II). There were no significant activities found with exceedingly high levels of the plant hormones gibberellic acid and 6-benzyladenosine (VII). There, also, was no significant larval growth reduction induced by the the commercial bioregulators glyphosate (VIII), maleic hydrazide and chlormequat chloride when these compounds were incorporated into diets at levels below those which normally caused phytotoxicity.

Table II. Antigrowth Activity of Several Plant Regulators to Corn Earworm.

Plant Regulator	Antigrowth Activity*		
Dinoseb (IX)	33% of control at	25	ppm
Glyphosate (VIII)	No significant effect at	150	"
Gibberellic acid (V)	" " " "	250	"
6-Benzyladenosine (VII)	" " " "	500	"
Maleic hydrazide (IV)	" " " "	500	"
Chlormequat chloride (II)	" " " "	1,000	"

*Corn earworm larvae grown for 12 days on a synthetic diet.

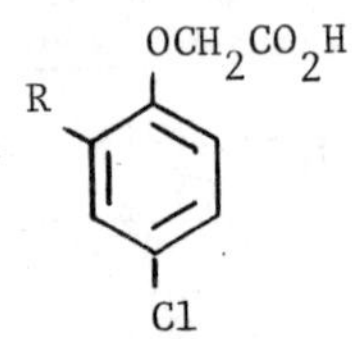

Ia, R=Cl
Ib, R=CH_3

$ClCH_2CH_2\overset{+}{N}(CH_3)_3$
Cl^-

II

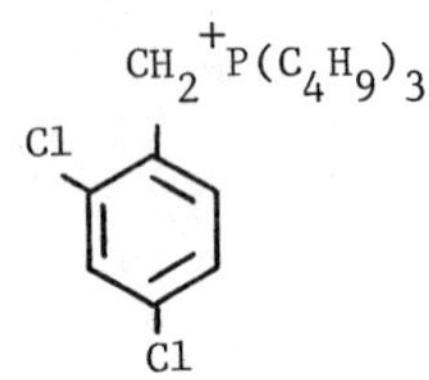

III

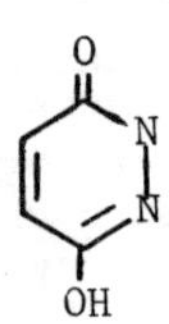

IV

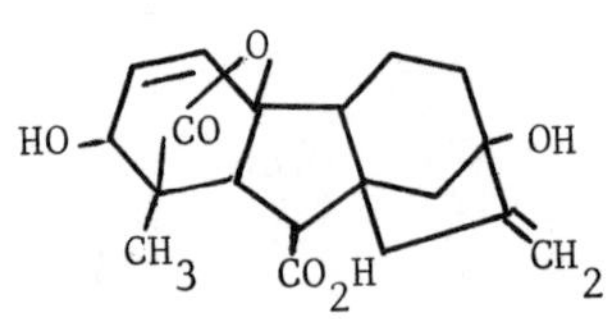

V

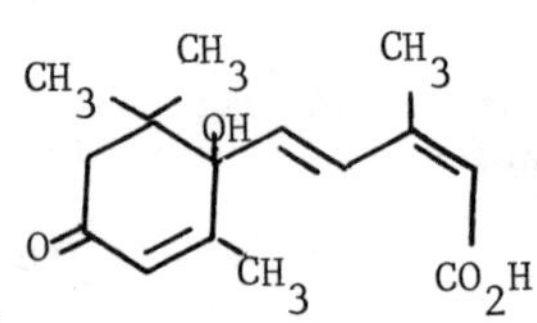

IV

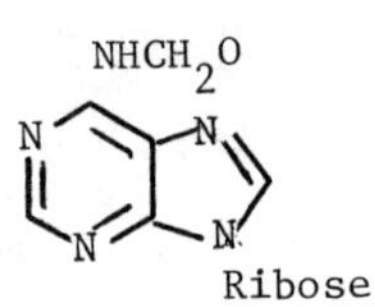

VII

$HN\ CH_2CO_2H$
|
$CH_2PO(OH)_2$

VIII

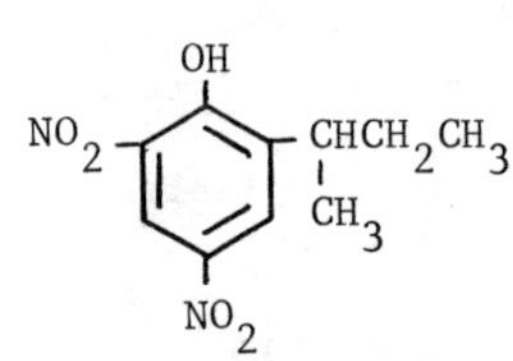

IX

$CH_2CONHN(CH_3)_2$
|
CH_2CO_2H

X

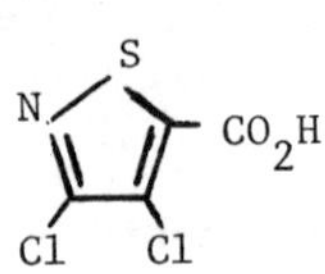

XI

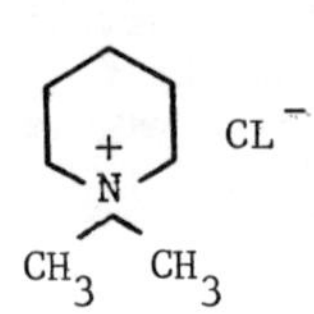

XII

Effect of Plant Regulators Treated Plants on Arthropods

Until the late 1950's the effect of herbicide and plant growth regulators on insect populations had been virtually ignored. A brief report on the effect of 2,4-D treated broad bean, Vicia fava L. on the reproductive rate of pea aphid was reported by Maxwell and Harwood in 1958 (26). They later suggested that the acceleration of aphid reproduction was due to an increase in free amino acid in the plant induced by the synthetic auxin (27). At about the same time, Robinson (28,29) reported that the application of the plant growth retardant, maleic hydrazide, increased the mortality of the nymphs and adults of the pea aphid on broad bean. However, he did not observe any effect on aphids grown on plants with 2,4-D and related growth stimulators. Later, however, Hintz and Schutz (30) showed that the application of several plant regulators, including 2-methyl-4-chlorophenoxy acetic acid (Ib), an analog of 2,4-D, unequivocally enhanced the reproduction of several aphid species grown on barley plants. The low levels of plant growth regulators used in Robinson's experiments may have been res responsible for the lack of observable change in aphid reproduction.

In the early 1960s several attempts were made to use the newly available plant growth regulator, chlormequat chloride (II) to modify and increase host plant resistance to pests and diseases. The results of these efforts, culminated in a symposium in 1968 were concerned mainly with the effect of chlormequat chloride and related growth retardant-treated plants on aphids, and spider mites (31-36,). It was suggested that the reduction in growth, development, survival, and fecundity of the aphids were due to the reduction of free amino acids in the treated plants (27,32,33). However, since aphids are phloem feeders, this view is difficult to substantiate until the concentration of free amino acids from the phloem, and not of the total plant, can be analyzed. The induced effects of plant regulator-treated plants to various sap-feeding insects and mites (38,38) are summarized in Table III.

Table III. Effect of Plant Regulator Treated Plants on Pest Populations.

Plant Regulator	Host plant	Pest	Pest Population	Ref.
(Ia)	Broad beans	Pea aphids	Increased	26,27
(IV)	" "	" "	Reduced	28,29
(Ib)	Barley	Aphids spp.	Increased	30
(II)	Brussel sprout	Cabbage aphid	Reduced	31-33
(II)	" "	Peach potato aphid	"	31,32
(II)	Broad beans	Bean aphid	"	33
(II), (III)	Chrysanthamum	Peach potato aphid	"	34
(II), (III)	Oleander	Oleander aphid	"	35
(II), (X)	Black current	Gall mite	"	36
(X)	Apple	Apple aphid and mite	"	37
(X)	Pear	Pear psylla	"	38

In addition to modifing the biochemical processes in the host plant, appropriately timed application of defoliants, desiccants, or growth terminating PGRs has been found to be a useful form of insect control. Kittock and co-workers (39,40,41) have found that chemical termination of late season vegetative and reproductive growth in cotton has a significant effect on the population of pink bollworm, Pectinophora gossypiella (Saunders). It appears that timely removal of immature cotton bolls with plant regulators such as chlormequat chloride and 3,4-dichloroisothiazole-5-carboxylic acid (XI) denies the late-season larvae adequate food to complete full pupation requisite for their over-wintering diapause. Similar techniques on cotton have been reported also for the possible control of the cotton bollworms (Heliothis zea and H. virescence) (42).

Recent Progress and Potential for Bioregulation of Host Plant Resistance

The effect of bioregulator treated sorghum on feeding behavior and fecundity of the greenbug, Schizaphis graminum (Rond), was recently investigated in our laboratory. Application of the plant regulators chlormequat chloride and mepiquat chloride (XII) to seedlings of greenbug-susceptible sorghum resulted in significant reduction in greenbug reproduction after the aphids were placed on mature plants 4 to 6 weeks later. However, there was no significant increase in inhibition on greenbug reproduction when the plant regulators were applied to greenbug-resistant lines. Electronic monitoring of aphid probing behavior (10) on chlormequat-treated, greenbug-susceptible sorghum showed a response pattern similar to that normally observed for greenbug probing on untreated resistant lines. It was demonstrated, for those aphid species which probe intercellularly and are phloem feeders, that intercellular pectin and related substances are the major barriers to aphid stylet penetration. The isolated pectin and its methoxy content of chlormequat treated, aphid-susceptible sorghum was over twice that of nontreated control plants.

Similar supression of aphid population on a chlormequat-treated, aphid-susceptible alfalfa line was also observed. Again there was no significant increase in inhibition of aphid reproduction when the bioregulators were applied to naturally aphid-resistant alfalfa plants. Attempts to correlate the polygalacturonase and esterase activities between greenbug biotypes with varying capacity to overcome resistance in sorghum are in progress.

Most of the work discussed above on host plant modification for insect resistance involves sucking insects but little has been reported in the literature on population suppression of two of the most damaging orders of chewing insects, the Lepidoptera and Coleoptera. The interaction between tomato plant and tomato fruitworm, H. zea was selected as a model system in our laboratory to demonstrate the feasibility of inducing resistance to chewing insects by means of plant bioregulators. This system was chosen because there is a diverse array of insect growth inhibitors and feeding deterrents already isolated and identified from the tomato plants (7,44). In addition, numerous studies and commercial applications of plant growth regulators on tomato have been reported, yet little is known regarding their effect on tomato fruitworm. Some of the uses of plant growth regulators on tomatoes includes the application of ethephon to accelerate fruit ripening and increase fruit coloration (45,46), 2-napthyl acidic acid or 2-chlorophennoxy acetic acid to improve fruit set and yield (47), absciscic acid to promote rooting of cutting (48), maleic hydrazide to prevent undesirable fertilization (49), daminozide to promote uniformed fruit ripening (50), and chlormequat chloride to increase resistance to Fusarium and frost hardiness (51,52).

Experiments in our laboratory have shown that growth of the tomato fruitworm on tomato plants treated with bioregulators such as glyphosate and chlormequat chloride was reduced significantly (Table IV). Mepiquat chloride at higher concentration, is also active against Heliothis larvae. Other growth regulators such as gibberellic acid, 6-benzyladenosine, dinoseb, 1-naphthyl acetic acid, ethephon and cycloheximide showed no dramatic activity in modifying the plant for larval growth suppression.

Table IV. Effect of Several Plant Regulator-Treated Tomato Plants on the Weight Gain of Tomato Fruitworm Larvae.

Plant Regulator	Conc. (ppm)	Ave. Larval Wt.* (mg)	% of Control
Control	--	451 ± 85	
Glyphosate	50	478 ± 42	106
	100	313 ± 124	69
	150	123 ± 72	27
Chlormequat ch.	25	404 ± 110	90
	250	284 ± 60	63
	1,000	167 ± 55	37
Gibberellic acid	10	554 ± 70	123
	50	448 ± 107	100
	100	580 ± 109	129

*Larvae grew from neonate for 12 days on detached foliage.

The biochemical mechanisms responsible for the induction of resistance H. zea in plant regulator -treated tomato plants is presently unknown. No significant correlation could be made between the reduction of larval growth and changes in levels of simple carbohydrates (e.g. fructose, glucose, sucrose, and inositol) or insect growth inhibitors (e.g. tomatine, rutin and chlorogenic acid) in tomato foliage. While both glyphosate and chlormequat chloride are generally regarded as plant growth retardants (53) their recognized modes of action are reported to be quite dissimilar (54,55). The diversity in the mode of action but similarity in the induction of insect resistance as well as the nature of the biochemical modification are being investigated.

Bioregulator induced resistance in corn against corn earworm, Heliothis zea (Boddie), should also be an excellent model system to demonstrate the usefulness of this novel pest suppression method. The period of vulnerability of corn to corn earworm is very short, approximately two weeks. Since young earworm larvae feed and survive mainly on fresh corn silks, promoting senescence or desiccation of corn silk with a plant bioregulator soon after fertilization of the ovule should reduce the food source and therefore the growth and survival of the insect. Our preliminary experiments show that treatment of corn plants 5 days after silks appear with 10 to 50 ppm of dinoseb, 2-sec-butyl-4,6-dinitrophenol (IX) results in a dramatic reduction in earworm growth to approximately one-half that of corn earworm reared on silk from untreated plant. There is further indication that the activity of the PBR depends much on the timing of application (i.e. age of plant) and genetic source of the plant. Similar observations were made when dinoseb was used to enhance the yield early maturity in corn (56).

As discussed earlier (Table II), there was no insecticidal effect from the various plant regulators on H. zea except for dinoseb at a level higher than was used in our experiments, hence, the activities of the plant regulator treatment on tomato and corn against H. zea appears to be a result of the biochemical modification of host plants.

For decades gossypol has been considered to be the insect resistance factor in cotton (57,58). Recently, condensed tannin and its monomeric flavonoid derivatives have been shown to be the factors important for the resistance of cotton to the cotton bollworms (H. virescence and H. zea) and pink bollworm, P. gossypiella (3,4). Because the flavonoid compounds and condensed tannin are derived from the phenylpropanoid biosynthetic pathway, the level of these compounds in cotton may be manipulated through bioregulation of phenylalanine ammonia-lyase (PAL), the key regulatory enzyme in this pathway (59-61). The complexity in bioregulation of allelochemicals in plants was revealed in our recent studies using L-amino-oxyphenylpropionic acid (L-AOPP), one of the most potent inhibitor of PAL, to manipulate the levels of phenylpropanoids in cotton (62,63).

When L-AOPP was fed at 10^{-3} M in 0.1 M sucrose solution, to cotton cotyledon, the apparent activity of PAL and the amount of condensed tannin was reduced from that of the control experiment. However, as the concentration of L-AOPP was decreased from 10^{-3} to 10^{-6} M, the activity of PAL was actually enhanced with a corresponding increased in the level of tannin, but that of the flavonoid gossypetrin, remained unchanged (64). It appears that the apparent activity of PAL is the net result of two opposing biochemical systems, the enzymatic synthesis and inactivation of PAL. In the present situation the PAL-inactivation system was more strongly inhibited by L-AOPP. As indicated by these experiments, bioregulation of secondary plant products can be both challenging and unexpected in a dynamic living system.

Concluding remarks

This review and the above initial experiments, hopefully, will serve to indicate the exciting challenges and potentials for pest suppression using plant bioregulators to manipulate the biochemistry and physiology of crop plants.

Because of a lack of experimental information, it is not possible, at present, to predict or generalize the influence of plant bioregulators have on host-pest interaction. Nevertheless an understanding of the compatibility between plant bioregulation and other methods of pest control is necessary for the effective use of plant regulators in an integrated pest management practices.

Success in manipulating host plant resistance through bioregulation should serve also to elucidate and/or re-affirm the mechanisms by which insect resistance is expressed in plants.

Reference to a company and/or product named by the Department is only for the purposes of information and does not imply approval or recommendation of the product to the exclusion of others which may also be suitable.

Literature Cited

1. Waiss, A. C., Jr.; Chan, B. G; Elliger, C. A. "Chemistry of Host Plant Resistance"; Hedin, P., Ed.; ACS Sympsoium Series, No. 66, American Chemical Society, Chicago, 1977, p. 115.
2. Waiss, A. C., Jr.; Chan, B. G.; Elliger, C. A.; Garrett, V. H.; Carlson, E. C.; Beard, B. Naturwissenchaften 1977, 64, 341.
3. Chan, B. G.; Waiss, A. C., Jr.; Lukefahr, M. J. Insect Physiol. 1977, 24, 113.
4. Chan, B. G.; Waiss, A. C., Jr.; Binder, R. G.; Elliger, C. A. Entomol. Exp. Appl. 1978, 24, 294.
5. Waiss, A. C., Jr.; Chan, B. G.; Elliger, C. A.; Wiseman, B. R.; McMillian, W. W.; Widstrom, N. W.; Zuber, M. S.; Keaster, A. J. J. Econ. Entomol. 1979,72, 256.
6. Dreyer, D. L.; Binder, R. B.; Chan, B. G.; Waiss, A. C., Jr.; Hartwig, G. L. Experientia 1979, 35, 1182.
7. Elliger, C. A.; Wong, Y.; Chan, B. G.; Waiss, A. C. Jr.; J. Chem. Ecol. 1981, 7, 753.
8. Dreyer, D. L.; Reese, J. C.; Jones, K. C.; ibid, 1981, 7, 2.

9. Waiss, A. C., Jr.; Chan, B. G.; Elliger, C. A.; Dreyer, D. L.; Binder, R. G.; Gueldner, R. C. Bull. Entomol. Soc. Amer. 1981, 27, 217.
10. Campbell, B. C.; McLean, D. L.; Kinsey, M. G.; Jones, K. C.; Dreyer, D. L. Entomol. Exp. Appl. 1982, 31, 140.
11. Maier, V. P.; Yokoyama, H. "Nutritional Qualities of Fruit and Vegetables". White, P. L., Ed., Futura, New York.
12. Ingram, J. W.; Bynum, E. K.; Charpentier, L. J. J. Econ. Entomol. 1947, 40 745.
13. Adams, J. B.; Drew, M. E. Can. J. Zool. 1965, 43, 789.
14. Fox, W. B. Sci. Agri. 1948, 28, 423.
15. Gall, A.; Dogger, J. R. J. Econ. Entomol. 1967, 60 75.
16. Leppick, E. E. Amer. Bee. J. 1951, 91, 462.
17. Johansen, C. Wash. Agric. Exp. Sta. Circ., 1959, 356.
18. El-Ibrashy, M. T.; Mansour, M. H. Experientia 1970, 26, 1095.
19. Tahori, A. S.; Zeidler, G.; Halevy, A. H. J. Sci. Fd. Agri. 1965, 16, 570.
20. Bhalla, O. P.; Robinson, A. G. J. Econ. Entomol. 1968, 61, 552.
21. Eichmeier, J.; Guyer, G. J. ibid, 1960, 53, 661.
22. Rodriguez, J. G.; Campbell, J. M. ibid, 1961, 54, 984.
23. Carlisle, D. B.; Ellis, P. E.; Osborne, D. J. J. Sci. Fd. Agric. 1969, 20, 391.
24. Visscher, S. N. Experientia 1980, 36, 130.
25. Spector, W. S., "Handbook of Toxicology"; Saunder, Philadelphia. 1956, 1, 118.
26. Maxwell, R. C.; Harwood, R. F. Bull. Entomol. Soc. Amer. 1958, 4, 100.
27. Maxwell, R. C.; Harwood, R. F.; Ann., Entomol Soc. Amer. 1960, 53, 199.
28. Robinson, A. G. Can. Entomol. 1959, 91, 527.
29. Robinson, A. G. Can. J. Plant Sci. 1961, 41, 413.
30. Hintz, S. D.; Schulz, J. T. Proc. N. Cent Branch - Entomol. Soc. Amer. 1969, 24, 114.
31. van Emden, H. F. Nature 1964, 201, 946.
32. van Emden, H. F. J. Sci. Fd. Agric. 1969, 20, 385.
33. Honeyborne, C. H. B. ibid, 1969, 20, 385.
34. Worthy, C. R. ibid, 1969, 20, 394.
35. Tahori, A. S.; Halevy, A. H.; Zeidler, G. ibid, 1965, 16, 568.
36. Smith, B. D. ibid, 1969 20, 398.
37. Hall, F. R. J. Econ. Entomol. 1972, 65, 1751.
38. Westigard, P. H.; Lombard, P. B.; Allen, R. B.; Strang, J. G. Environ. Entomol. 1980, 9, 275.
39. Kittock, D. L.; Manney, J. R.; Arle, H. F.; Bariola, L. A. J. Environ. Quality 1973, 2, 405.
40. Bariola, L. A.; Kittock, D. L; Arle, H. F.; Vail, P. V.; Henneberry, T. J. J. Econ. Entomol. 1976, 69, 633.
41. Kittock, D. L; Arle, H. F.; Henneberry, T. J.; Bariola, L. A.; Walhood, V. T. Crop Sci. 1980, 20, 330.

42. Thomas, R. O.; Cleveland, T. C.; Cathey, G. W. Crop Sci. 1979, 19, 861.
43. Dreyer, D. L.; and Campbell, B. C. Experientia (in press).
44. Williams, W. G.; Kennedy, G. G.; Yamamoto, R. T.; Thacker, J. D.; Bordner, J. Science 1980, 207, 888.
45. Dostal, G. H.; Wilcox, G. E. J. Amer. Soc. Hort. Sci. 1971, 96, 656.
46. Buescher, R. W. Hort. Science 1977, 12, 315.
47. Lukasik, S. Acta Agrobot. 1975, 28, 43.
48. Basu, R. N.; Roy, B. N.; Bose, T. K. Plant Cell Physiol. 1970, 11, 681.
49. Rehm, S. Nature 1952, 170, 38.
50. Samish, R. M. Ann Rev. Plant Physiol. 1954, 5, 183.
51. Davis, D.; Diamon, A. E. Phytopath. 1953, 53, 137.
52. Bierecka, H.; Zebrowski, Z. Bull. Acad. Pol. Sci. 1966, 14, 367.
53. Anderson, A. S. "Plant Regulator and Word Agriculture", Scott, T. K., Ed., 1979, Plenum.
54. Dicks, J. W. "Recent Development in the Use of Plant Growth Retardants"; Clifford, D. R. and Lenton, J. R., Eds; British Plant Growth Regulator Group Symposium, London, 1979, p. 1.
55. Hoagland, R. E.; and Duke, S. O. "Biochemical Responses Induced by Herbicides", Moreland, D. E.; St. John, J.; Hess, P. D.,Eds., Amer. Chem. Soc., Symposium Series 181, 1982, 175.
56. Hatley, O. E.; Hermann, L.; Ohlrogges, A. J. Proc. Plant Growth Reg. Work Group", 1977, 4, 156.
57. Quaintance, A. L.; Brues, C. T.; U.S. Department of Agricultural Bur. Entomol. Bull., 1905, 155.
58. Bottger, G. T.; Sheenhan, E. T.; Lukefahr, M. J. J. Econ. Entomol. 1964, 57, 283.
59. Hahbrock, K. Eur. J. Biochem. 1976, 63, 137.
60. Luckner, M. "Encylopedia of Plant Physiology" Bell, E. A. and Chardwood, R. V., Eds., 1980, 8, 24, Springer-Verlag, N.Y.
61. Creasy, L. L.; and Zucker, M. "Recent Advances in Phytochemistry" Runeckles, V. C.; and Conn, E. E., Eds., 1974, 8, 1, Academic Press, N.Y.
62. Hanson, K. R. Arch. Biochem. and Biophys. 1981, 211, 564.
63. Amrhein, N.; Gödeke, K. H. Plant Sci. Lett. 1977, 8, 313.
64. Creasy, L. L.; Chan, B. G. J. Econ. Entomol. 1984.

RECEIVED March 19, 1984

Effects of 1,1-Dimethylpiperidinium Chloride on Growth and Water Relations of Cotton in a Semiarid Environment

C. W. WENDT, V. R. ISBELL, B. L. STUART, and J. R. ABERNATHY

Texas Agricultural Experiment Station, Lubbock, TX 79401

Application of mepiquat chloride (1,1-dimethyl-piperidinium chloride) to cotton grown on the Southern High Plains of Texas in 1980 resulted in reduced plant growth and increased leaf water potential components and transpiration rate, with no decrease in lint yield. These results indicate that this bioregulator improves plant water status. Leaf area reductions of 27 percent, water potential increases of 0.20 to 0.45 megaPascals (MPa), and 13 to 28 percent higher transpiration rates were measured in mepiquat chloride treated plants. No differences in the amount of soil water extracted were measured as a result of treatment with mepiquat chloride.

The semi-arid areas of the world are among the last frontiers in which agricultural production may be increased. They present a major challenge to establishment of stable production in that the climate, soils, water supplies, and cropping systems are quite variable. The Southern High Plains of Texas is typical of such a region. Although annual precipitation averages approximately 450 mm, it may vary from 225 mm to 1000 mm. Monthly precipitation extremes range from 0 to 350 mm. Two-thirds of the precipitation occurs between May and September. Windspeeds are high, averaging 6 m/sec throughout the summer growing season. Winds greater than 11 m/sec often continue for periods of more than 12 hours. The average annual temperature is 16° C, with July being the warmest month with an average high of 33° C. Relative humidity is generally low, averaging approximately 36 percent at late afternoon, during the summer. Annual evaporative demand is approximately 2000 mm which exceeds precipitation by 400 percent. Cool temperatures during both the spring and fall often limit the growing season of some summer-grown crops.

Approximately 50 percent of the area is irrigated from the Ogallala aquifer formation, a diminishing water supply. The remaining agricultural area is rainfed or dryland. The percentage of dryland agriculture is increasing because of diminishing water supplies and/or high energy costs.

0097-6156/84/0257-0205$06.00/0

Soils in the area vary in texture from sands to clays. Cotton, grain sorghum, wheat, and corn are the major crops of the area. Cotton is the dominant crop in the southern portion of the region. It is in this region that irrigation has been in practice the longest, and consequently underground water supplies are lowest there.

Overall, one of the most crucial limitations in crop production is availability of water. In response to this limitation various research approaches have been initiated. These include breeding for drought tolerance and various management techniques such as antitranspirants, soil water evaporation suppressants, more efficient irrigation systems, crop rotations, soil surface modifications, and optimization of soil fertility-water combinations.

Water stress causes a variety of complex, interdependent plant responses. There has been limited research on the use of bioregulators to enhance plant production in situations of plant water stress. The effect of many synthetic compounds on transpiration rates in a variety of crops has been reported (1-4). Recently, attention has been directed to the relationship between plant water stress and the activity of endogenous hormones (5-9). The ability of bioregulators to alter the activity of these hormones has been postulated, and in some cases demonstrated (10).

In 1979 we began to evaluate the possibility of using a plant growth regulator, mepiquat chloride (1,1-dimethyl-piperidinium chloride), on cotton. At that time reports of mepiquat chloride (MC) effects on cotton growth and production in the humid areas of Mississippi, Louisiana, and Arkansas, as well as the irrigated production regions in Arizona, California, and the Rio Grande Valley of Texas were being circulated (11-13). The most commonly reported effects of MC on vegetative growth of cotton have been decreases in plant height and leaf area (11,14,15). The resulting modification of canopy architecture has resulted in improved lint yield largely because of reduced boll rot and increased efficacy of cultural inputs such as insecticides (14,16). Effectiveness of MC in control of excess vegetative growth has led researchers to evaluate its potential in production systems which inadvertantly are conducive to development of nonproductive vegetative growth. These studies have indicated that the deleterious effects on lint maturity and yield that may occur from high soil nitrogen (17-19), high soil water availability (17,19,20), and high plant populations (19,20), may be lessened by application of MC. These studies were of limited value in the Texas High Plains. Cotton grown here is naturally smaller-statured (leaf area index < 3) because of genetic or environmental growth limitations. Consequently, problems associated with excessive vegetative growth are encountered only occasionally.

Our investigations in the Texas High Plains have been directed to the possibility of improvement of plant water relations in response to MC. A key to attaining this goal may be the exploitation of MC effects on the regulation of leaf area development and leaf structure. Mepiquat chloride has been reported to increase leaf thickness and alter cell structure within the leaf (15,21,22). Modification of leaf blade morphology has been related to improved leaf water status in cotton (23). The reported effects of MC may induce such an improvement.

The management of leaf area development by MC may allow an increase in the root:shoot ratio, if root growth is relatively insensitive to the growth regulator. In theory, this situation would improve water availability to the shoot as water stress conditions develop, and allow soil water to be distributed over a longer period. This modification could increase economic yield by decreasing the severity of water stress during the critical period of lint development. Modification of plant water status suggested the potential for enhancement of production efficiency in our semi-arid environment.

Materials and Methods

An evaluation of the effects of MC on plant growth, development, lint yield, and water relations was conducted from 1979 to 1981 at the Texas Agricultural Experiment Station near Lubbock, Texas. MC was applied to the Paymaster 303 cultivar during early square stage (young floral bud) at a broadcast rate of 50 g ai/ha. The soil type was an Olton loam soil (fine, mixed, thermic family of Aridic Palenstolls). Data was collected from replicated field experiments involving a range of environmental conditions and cultural practices. The detailed presentation of these experiments is in press (24). Plant growth measurements made during the season included height, leaf area index (LAI), and shoot dry matter accumulation. Plant water relations were monitored by measuring leaf water potential components, stomatal diffusive resistances, and transpiration rates during diurnal cycles, and throughout the season. Leaf water potential components were determined using leaf cutter thermocouple psychrometers as previously described (24). Abaxial transpiration rate, diffusive resistance, and leaf temperature were measured with a steady state porometer. Soil water content was measured throughout each season using a neutron probe at 0.3 m vertical intervals to a depth of 1.8 m. Lint and seed yields were recorded and lint quality determined. The data presented herein are from a representative field experiment conducted in 1980. The study was planted with Paymaster 303 to a final population of 21 plants/m^2. During this season soil water conditions were considered to be suboptimal for maximum dry matter production (a preplant and one furrow irrigation 44 days after planting). The growing season of 1980 was typified by favorable conditions during stand establishment, followed by extremely hot, dry conditions through mid-August.

Results

Shoot height was significantly reduced in MC treated plants at 17 days after application (DAA) through 74 DAA (Figure 1a). Inhibition of shoot extension was the result of a reduction in the number of mainstem nodes and/or internode length (data not shown). Total shoot dry matter accumulation was slightly reduced in the treated plants only at 31 DAA (Figure 1a). Leaf area index (LAI) was reduced 27 percent in treated plants 31 DAA (Figure 1b). Other experiments indicated that LAI reductions were a function of both reduced leaf number and reduced leaf blade area (24). Across all experiments the ability of MC to inhibit growth was related to environmental constraints. Under conditions of severe water stress the growth of plants was restricted irrespective of growth regulator treatment.

Alterations in daily minimum leaf water potential components were evident soon after MC application. The total water potential

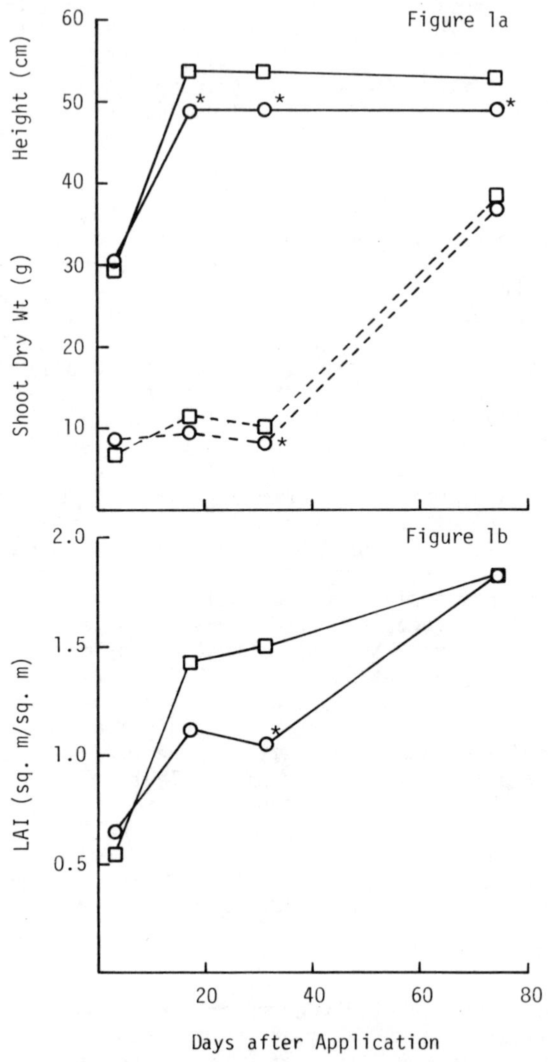

Figure 1. Plant height (Figure 1a solid lines), total shoot dry weight (Figure 1a dashed lines), and leaf area index (Figure 1b) of the control (squares) and MC (circles) treatments on 3, 17, 31, and 74 days after application. Asterisks indicate significant difference at P = 0.05.

(Ψw) was 0.2 megaPascals (MPa) higher in the MC treated plants 14 DAA (Figure 2a). Subsequent measurements on 18, 26, and 49 DAA also showed higher Ψw in treated plants with a maximum increase of 0.45 MPa, 49 DAA. Differences in solute potential (Ψs) at daily minimum Ψw were not as consistent statistically as Ψw; however, Ψs values were generally higher in treated plants (Figure 2b).

Figures 3a and 3b present the diurnal changes in Ψw and Ψs as influenced by treatment. Measurement of Ψw and Ψs at 0600 hours indicated that MC treated plants maintained water status comparable to untreated plants prior to the beginning of a dehydration cycle. Ψw values between treatments diverged as the day progressed. Increased Ψs in treated plants was also found; however, the increase in Ψs was not as great as the increase in Ψw. In this experiment, as in others, Ψw differences became more pronounced late in the afternoon. Higher Ψs (and calculated turgor pressure) in treated plants was more common during the morning than the afternoon.

Abaxial transpiration rates were measured on three dates at daily minimum Ψw in this experiment. On each date the transpiration rate of treated plants was higher than that of untreated plants with increases ranging from 13 to 28 percent (Figure 4).

No significant differences between treated and untreated plants were measured in lint yield, lint quality, or seasonal soil water use (data not shown).

Conclusions

Reductions in seasonal soil water extraction from MC modification of leaf area development were not demonstrated even in experiments conducted under high soil water levels. Partial explanation of this finding is provided by the measurement of increased transpiration rates in treated plants when LAI was reduced.

Improvements in plant water status (higher Ψw, Ψs, and Ψp) were consistently found in this experiment and others regardless of whether or not MC affected measured growth parameters. The potential for combining some of the positive physiological modifications by MC with additional complementary changes (e.g. optimization of photosynthesis-transpiration relationships) provides future research opportunities.

Our inability to demonstrate differences in lint yield or soil water use in cultural conditions considered representative for this area does not preclude the utility of MC in production systems which predictably cause excess vegetative growth.

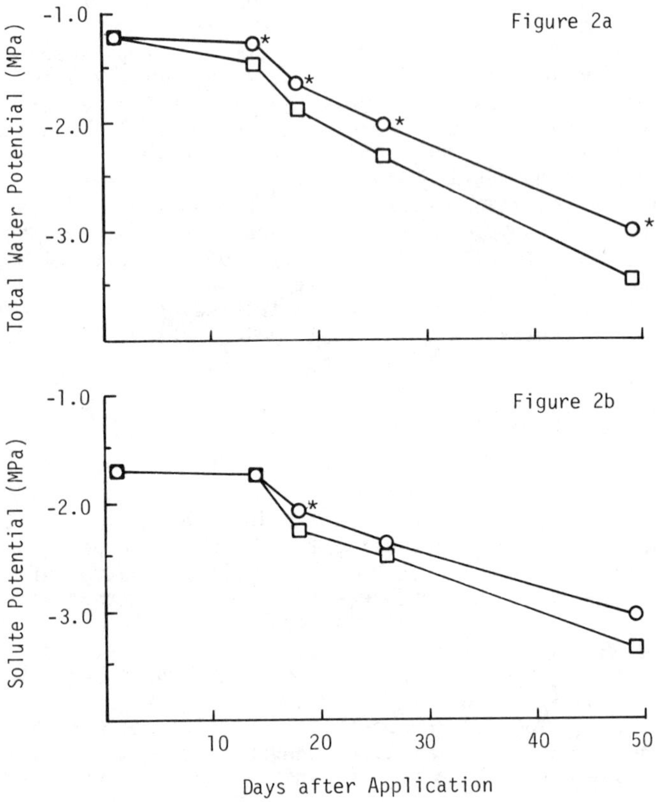

Figure 2. Total leaf water potential (Figure 2a) and solute potential on 1, 14, 18, 26, and 49 days after application (squares--control; circles--MC). Asterisks indicate significant difference at P = 0.05.

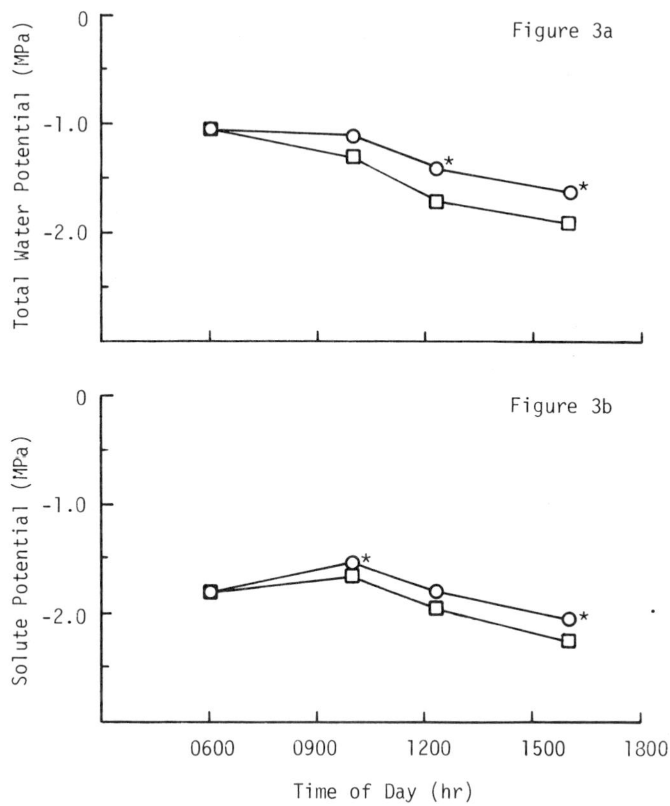

Figure 3. Diurnal variation in total leaf water potential (Figure 3a) and solute potential (Figure 3b) at 18 days after application in the control (squares) and MC (circles) treatments. Asterisks indicate significant difference at P = 0.05.

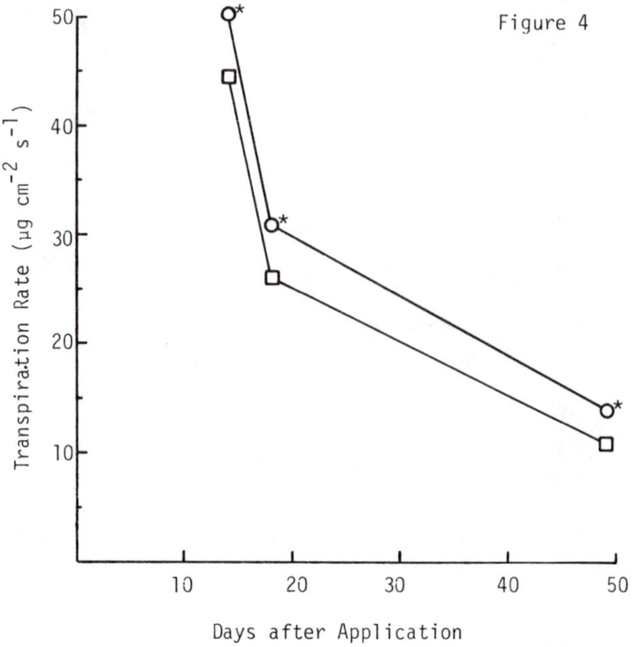

Figure 4. Abaxial transpiration rate of the control (squares) and MC (circles) treatments at daily minimum leaf water potential on 14, 18, and 49 days after application. Asterisks indicate significant difference at $P = 0.05$.

Literature Cited

1. Slatyer, R.O.; Bierkuizen J.F. Agri. Meteorol. 1964, 1, 42-53.
2. Gale, J.; Hagan, R.M. Ann. Rev. Plant Physiol. 1966, 17, 269-282.
3. Santakumari, M.; Reddy, C.S.; Rama Dos, V.S. Proc. Indian Acad. Sci. 1977, 86B, 143-150.
4. Ahmed, S.; Fletcher, R.A. Weed Sci. 1980, 28, 180-185.
5. Livine, A.; Vaadia, Y. In "Water Deficits and Plant Growth." Vol. 3; Kozlowski, T.T., Ed.; Academic Press, New York. 1972. Chap. 8.
6. Vaadia, Y. Phil. Trans. R. Soc. Lond. B. 1976, 273, 513-522.
7. Itai, C.; Benzioni, A. In "Water and Plant Life: Problems and Modern Approaches"; Lange, O.L.; Kappen, L.; Schulze, E.-D., Eds.; Springer-Verlag: Berlin; Heidelberg; New York, 1976; Chap. 4B.
8. Aspinall, D. In "Adaptation of Plants to Water and High Temperature Stress"; Turner, N.C.; Kramer, P.J., Eds.; Wiley and Sons; New York, 1980; Chap. 11.

9. Simpson, G.M. "Water Stress On Plants"; Praeger, New York, 1981; pp. 127-139.
10. Norman, S.M.; Poling, S.M.; Maier, V.P.; Orme, E.D. Plant Physiology. 1983, 71, 15-18.
11. Willard, J.I.; Thompson, J.T.; Daniel, J.W.; Ware, T.; Steinert, W.G.; Carter, C.W.; Schott, P.E. Proc. 4th Ann. Mtg. Plant Gro. Reg. Work Grp., 1977, p. 341.
12. Namken, L.N.; Gausman, H.W. Proc. Beltwide Cotton Prod. Mech. Conf., 1978, p. 23-25.
13. Erwin, D.C.; Isai, S.D.; Khan, R.A. Phytopathology 1979, 69, 283-287.
14. Gausman, H.W.; Namken, L.N.; Herlman, M.D.; Walter, H.; Ritting, F.R. Proc. Beltwide Cotton Prod. Res. Conf., 1979, p.51.
15. Walter, H.; Gausman, H.W.; Rittig, F.R.; Namken, L.N.; Escobar, D.E.; Rodriguez, R.R. Proc. Beltwide Cotton Prod. Res. Conf., 1980, pp. 32-25.
16. Chambers, A.Y.; McCutchen, T.C. Proc. Beltwide Cotton Prod. Res. Conf., 1977, pp. 28-29.
17. Heilman, M.D. Proc. Beltwide Cotton Prod. Res. Conf., 1981, p. 47.
18. Kerby, T.A.; George, A.; Hake, K.; McCutcheon, O.D.; Vargas, R.N.; Weir, B.; Brittan, K.; Kukas, R. Proc. Beltwide Cotton Prod. Res. Conf., 1982, p. 54.
19. York, A.C. Agron. J 1983, 75, 667-672.
20. Briggs, R.E. Proc. Beltwide Cotton Prod. Res. Conf., 1980, p. 32.
21. Gausman, H.W., Ritting, F.R.; Namken, L.N. Proc. Beltwide Cotton Prod. Res. Conf., 1978, p. 40.
22. Gausman, H.W.; Ritting, F.R.; Namken, L.N.; Rodriques, R.R.; Escobar, D.E.; Garza, M.V. Proc. 5th Ann Mtg. Plant Gro. Reg. Work Grp., 1978, pp. 137-145.
23. Karami, E.; Krieg, D.R.; Quisenberry, J.E. Crop Sci. 1980, 20, 421-426.
24. Stuart, B.L.; Isbell, V.R.; Wendt, C.W.; Abernathy, J.R. Agron, J. (In Press).

RECEIVED January 31, 1984

18

Use of Leaf Optical Properties in Plant Stress Research

HAROLD W. GAUSMAN, JOHN J. BURKE, and JERRY E. QUISENBERRY

Plant Stress and Water Conservation Research Unit, Agricultural Research Service, U.S. Department of Agriculture, Texas Tech University, Lubbock, TX 79409

Leaf reflectance may become a "diagnostic tool" to detect plant stress in three areas: chlorophyll absorption (0.45- to 0.75- μm waveband), near-infrared (0.75- to 1.35- μm waveband), and infrared-water absorption (1.35- to 2.5- μm waveband) spectral regions. Stress may either increase or decrease leaf total chlorophyll concentration: low chlorophyll concentrations usually cause a higher reflectance at 0.55- μm than higher concentrations. Furthermore, stress inhibts leaf expansion. A stressed leaf's mesophyll remains compact and consequently has a lower reflectance than a "nonstressed" leaf that has a lacunose mesophyll. The higher near-infrared light reflectance for a nonstressed leaf is caused by more light scattering, when light goes from hydrated cell walls (refractive index of 1.4) into airspaces (refractive index of 1.0). Moreover, water absorbs infrared light energy within the 1.35- to 2.5- μm waveband: leaves with high water content have lower reflectance than do leaves with lower water content. Practical aspects are reviewed.

Effects of stresses on plants impact on the world's crop production, particularly stresses from extremes of temperature and water availability. Therefore, attempts should be made to improve the water-use efficiency and temperature tolerance of plants. To this end, techniques to detect plant stress are needed so that stress resistant germplasm and water conserving management practices can be developed.

When a leaf intercepts incoming radiation (light) at a critical angle, a portion of the light is absorbed. The amount of absorption depends on the energy (wavelength) of the photons involved, photons with the shortest wavelength (high energy ultraviolet and visible light) are involved in photosynthesis and photochemical reactions as after such photomorphogensis, phototropism, and chlorophyll synthesis; whereas, photons with the longest wavelength (low energy infrared light) affect heating processes, evaporation, and transpiration. Consequently, changes in chloroplast numbers and in the concentration of leaf chloroplasts' pigments and tissue water content are largely responsible for inducing variability in plant tissue light absorption.

A portion of the incoming photons are also reflected when they impinge on a leaf at a critical angle. Specular reflectance occurs at the leaf cuticle; whereas, diffuse reflectance originates from light scattering (multiple reflection) mainly within the leaf mesophyll. Light that is neither absorbed or reflected is transmitted. In a plant canopy, therefore, transmitted light usually interacts with subtended leaves or with the soil, giving many ramifications to the understanding of spectral characteristics of diverse types of plant canopies.

Most spectral measurements reported on here were made over the 0.5- to 2.5- μm waveband on single plant leaves in the laboratory by using a spectrophotometer with a reflectance attachment. To facilitate interpretation of data the 0.5- to 2.5- μm waveband, which includes most of the visible light region (0.4- to 0.75- μm) and the infrared region (0.75- to 2.5- μm) has been divided into three categories. These are: (1) the 0.5- to 0.75- μm visible light absorptance region, dominated by pigments -- primarily chlorophylls *a* and *b* (chlorophyll *c* occurs in brown algae), carotenes, and xanthophylls; (2) the 0.75- to 1.35- μm near-infrared waveband, a region of high light reflectance and low absorptance that is affected considerably by internal leaf structure; and (3) the 1.35- to 2.5- μm waveband, a region influenced by some leaf structure, but affected greatly by water concentration in the tissue -- strong water absorption bands occur at the 1.45- and 1.95- μm wavelengths. The peaks following these bands at 1.65- and 2.2- μm wavelengths, respectively, decrease as leaf tissue water content increases.

It is extremely important to consider leaf maturity if spectrophotometric measurements are to be made to compare the reflectance of stressed with nonstressed leaves. For example, cotton (*Gossypium hirsutum*) leaves were sampled fourth from the apex of salinity-stressed and nonstressed plants (*2*). Spectrophotometric measurements showed that salinity-stressed leaves had higher reflectance than nonstressed leaves. However, nonstressed plants were growing more rapidly than stunted, salinity-stressed plants. Hence, leaves fourth from the apex of nonstressed plants were younger or more compact than more porous leaves fourth from the apex of stressed plants. When leaves were sampled at the same chronological age, stressed leaves had lower reflectance than nonstressed leaves because stressed leaves were stunted or more compact than the more porous, nonstressed leaves of the same chronological age. Thus, when leaf maturation was not considered, stressed leaves had higher reflectance than did nonstressed leaves; if leaf maturation was considered, stressed leaves had lower reflectance than did nonstressed leaves.

Several factors (stresses) that affect the leaf's optical properties, mainly light reflectance measurements, will be considered below. Some attention, however, will be given to newer techniques such as fluorescence determination. For simplicity the term "light" will be used instead of electromagnetic radiation, and it will include both visible and near-infrared light spectra.

Water and Temperature

Water and temperature stresses are insidious culprits and have been included in the same section because it is very difficult, if not impossible, to separate their effects on plant growth under field conditions with an extremely variable environment. For example, as plant leaf temperature increases, transpiration generally decreases. As summarized by Nixon *et al.* (3), plant temperatures increased with decreasing availability of soil moisture. For example, plant canopy differences up to 6 C existed between the most and the least water-stressed cotton fields. This temperature difference was very easy to detect with various remote sensing instruments in the thermal infrared spectral region, especially at a waveband encompassing the 10.3- μm wavelength.

A method to detect freeze injury has been reported (4). Frozen sour orange (*Citrus aurantium*) leaves had lower reflectance over the 0.5- to 2.5- μm waveband than nonfrozen leaves because freezing apparently destroyed the semipermeability of their cell membranes; subsequently, intercellular sap migrated to intercellular spaces. Filling of air spaces with sap decreased the number of light-scattering, hydrated cell wall-air interfaces; consequently, reflectance was decreased. Reflectance measurements showed statistically significant differences between noninjured and freeze-injured leaves, regardless of their watersoaked or nonwater-soaked appearance. It was concluded that reflectance measurements should be useful to detect cell membrane leakage, supposedly injury, in citrus leaves.

Effects of drought and/or temperature stress can be

evaluated within the 1.35- to 2.5- µm waveband water-absorption region. Wavebands around the 1.65- to 2.2- µm wavelengths, for example, can be used to evaluate leaf water content (5), and detect general leaf freeze damage (6).

Leaf dehydration greatly increases the spectrophotometrically measured light reflectance over the 0.5- to 2.5- µm waveband. Thomas et al., (9) found that reflectance increased as relative turgidity (10) decreased below values of 80 percent at selected 0.54-, 0.85-, 0.65-, and 1.45- µm wavelengths. Relative turgidity is used to measure plant water stress. It is the actual leaf water content expressed as a percentage of the turgid or saturation water content. Regression equations were calculated (11) to express the incident light reflectance from the upper (adaxial) single leaf surfaces as a function of relative turgidity and water content. Reflectances at the 1.45- and 1.95- µm wavelengths were related to the leaf relative turgidity or relative water content. However, because of variations in internal leaf structure that was apparently associated with water availability during leaf development, the ability to predict leaf water status from reflectance measurements was poor. With cotton (Gossypium hirsutum), the greatest reflectance change occurred when the relative turgidity was below 70 percent, and the leaves were visibly wilted. Within the relative turgidity range from 70 to 80 percent, reflectance changes were small, and they were not always definable for predictive purposes because of variation among leaves of field-grown cotton plants, caused by age (maturation) differences and osmotic stresses. Carlson et al., (12) found that corn (Zea mays), grain sorghum (Sorghum bicolor), and soybean (Glycine max) leaf reflectances were highly linearly correlated with relative water content at two strong water-absorbing wavelengths, 1.45 and 1.95 µm, and two wavelengths of lower absorptivity, 1.1 and 2.2 µm.

As alluded to previously, it is usually very difficult, if not impossible to distinguish between temperature and water stress effects on the reflectance of single plant leaves because temperature and water relations are so closely related.

Soil Salinity

Morphological studies indicated that plant species from saline environments had thicker leaves, more developed palisade parenchyma, fewer chloroplasts, smaller intercellular spaces, and fewer stomata per unit area (13, 14, 15, 16, 17) than plants from nonsaline areas. Salinization also reduced cell division in leaves of cotton (Gossypium hirsutum) (18), and reduced the rate of cell enlargement and of protein and ribonucleic acid (RNA) synthesis in bean leaves (Phaseolus vulgaris) (19).

Spectrophotometric studies showed that leaves of cotton plants affected by soil salinity had reduced light reflectance and increased light transmittance as compared with that of unaffected leaves of the same chronological age (20). The salinity-stressed leaves were stunted with a more compact cell arrangement than nonstressed leaves. Cotton plants affected by soil salinity appeared darker red on Kodak Ektachrome Infrared Aero 8443, color infrared transparencies and prints as compared with those of "normal"-appearing plants. This was caused by a higher chlorophyll concentration in high-salt than in low-salt leaves. Leaves with high chlorophyll content induced a darker red magenta tone than did leaves with low chlorophyll; because high chlorophyll increased red light absorptance, decreased its reflectance (less radiation impinging on the film), and caused a more saturated image in the magenta dye layer, which allowed less green light transmittance, and thus, produced a darker appearance. Less chlorophyll caused higher red light reflectance, less magenta dye, and lighter appearance. To understand this, the basics of color-infrared film must be reviewed. The color-infrared film has three image layers individually sensitized to green, 0.5- to 0.6- μm waveband; red, 0.6- to 0.7- μm waveband; and near-infrared light, 0.7- to 0.9- μm waveband, instead of to blue, green, and red light for Ektachrome noninfrared film (21). A yellow filter is used on the camera with color infrared film to absorb the blue light, to which these layers are also sensitive. Upon processing, yellow, magenta, and cyan positive images are formed in the green-, red-, and near-infrared-sensitive layers, respectively. The overall impression to an observer viewing the finished print or transparency will depend upon which one of the three positive images in the dye layers predominate with respective to visual appearance.

Because the eye sensitivity peaks in the green, the magenta layer generally contributes most to the subjective impression of lightness or darkness in a color print or transparency. For example, healthy leaves, with high near-infrared, as compared with those of low near-infrared reflectance for unhealthy leaves, record red because a light-toned cyan image (less dense or less saturated) results, which allows the transmittance of more red radiation in the viewing.

Salinity-affected areas are usually easy to detect in cotton fields, although it is sometimes difficult to distinguish saline areas from cotton root rot (Phymatotrichum omnivorum) areas (22). Generally, on near-infrared color film, healthy cotton plants appear to be magenta, soil saline areas are whitish, and root rot-affected areas are darker blotches with a sharper demarcation around their perimeters.

Nutrient Deficiency or Toxicity

A spectrophotometer was used to measure diffuse reflectance from upper (adaxial) leaf surfaces of sweet pepper (*Capsicum annum*), cabbage (*Brassica oleracea*), and spinach (*Spinacia oleracea*) (23) leaves to quickly estimate their nitrogen status. Leaf light absorptance in the visible region is primarily dependent on chlorophylls *a* and *b* and carotenoid (carotene and xanthophyll) concentrations in components (grana) of the chloroplasts. Green leaves absorb 75 to 90 percent of blue light (about the 0.45- μm wavelength) and red light (about the 0.68 μm wavelength) of the visible spectrum. Absorptance is smallest in the waveband around the 0.55 μm wavelength, where a green light reflectance peak of usually less than 20 percent occurs. Since nitrogen nutrition of plants markedly affects pigment concentrations and subsequent leaf color, it was surmised that limiting the amount of nitrogen would reduce pigment concentrations and therefore increase reflectance because of decreased radiation absorptance. This tenet was supported by research: reflectance was inversely correlated with the leaf nitrogen content of three plant species. Regression equations that were developed expressed reflectance as a function of leaf nitrogen content of greenhouse-grown plants. These functions can be used to estimate the nitrogen content of field crops. With field-grown sweet peppers, for example, the difference between Kjeldahl-determined and reflectance-estimated nitrogen content was less than 0.7 percent.

Remote sensing techniques may be used to determine some nutrient difficiencies. They have been based on or instigated by spectrophotometric single leaf measurements conducted in the laboratory. Such was the case with reflectance differences between chlorotic (iron deficient) and green ("normal") grain sorghum (*Sorghum bicolor*) plants. For example, multispectral data from the ERTS-1 satellite detected differences in chlorophyll concentration between chlorotic and green grain sorghum plants (24). Band 5 (0.6- to 0.7- μm waveband) data were used, representing the chlorophyll absorption band at the 0.65-μm wavelength. Chlorotic sorghum areas 2.8 acres (1.1 hectares) or larger were identified on a band 5 computer printout. However, it is very difficult and quite often impossible to distinguish among the single leaf spectra for several nutrient deficiencies (25), or among the single leaf spectra for nutrient deficiencies and a spectrum for a nutrient toxicity (26).

Plant Maladies

There are numerous reports in the literature that indicate the usefulness of infrared film or spectrophotometric measurements to detect plant maladies: diseases, nematode damage, insect damage, etc. However, only a few of the most classical examples will be considered below.

Colwell (27) was one of the first investigators to suggest using infrared film to record any disease that interfered with the internal reflection of light within leaves. Keegan et al., (28) studied effects of stem rust (Puccinia graminis tritici) and leaf rust (Puccinia triticina or Puccinia rubigovera tritici) of wheat on light reflectance. The data showed that severe, as compared with that of low rust infestation, caused a rounding of the near-infrared reflectance plateau's shoulder over the 1.0- to 0.75- μm waveband. A similar reflectance response was noted after hair removal on upper (adaxial) leaf surfaces of the velvet plant (Gynura aurantiaca) (29).

Cellular discoloration within leaves may be useful in detecting nonvisual symptoms of plant maladies. Often, this is referred to as "previsual detection", which is considered to be a misnomer by many scientists. Manzer and Cooper (30) found that potato (Solanum tuberosum) late blight (Phytophthora infestans) could be detected by aerial photography from three to five days before visual symptoms became apparent; tobacco (Nicotiana tabacum) ringspot virus (Nicotinia spp. virus) could be detected about one day before visual symptoms were evident (31); and ozone-damaged leaf areas of cantaloupe (Cucumis melo) plants were detected photographically 16 hours before the damage was visible (32). Conversely, Heller (33) found that beetle damage could not be predicted, and Meyer (34) reported that too much biological variability interfered with previsual detection of tree disease.

To study the effects of internal leaf damage on light energy spectra, cotton (Gossypium hirsutum) leaves were infiltrated with anhydrous ammonia, (35). Spectrophotometric laboratory measurements on anhydrous ammonia-treated leaves showed more absorptance and less reflectance and transmittance than untreated leaves over the 0.75- to 1.35- μm waveband and the 0.50- to 0.75- μm waveband. Apparently, the brownish discoloration caused by the anhydrous ammonia increased leaf opaqueness.

Reflectance was reduced by severe rust infection on Westar wheat (Triticum aestivum) leaves (36), benzene vapor on cotton leaves (unpublished), natural freezing of sea grape (Coccoloba uvifera) leaves (unpublished), ammonia treatment of cotton leaves, hair removal by shaving of velvet plant leaves (37), and leaf freeze injury (38).

One of the most successful and unique examples in applying laboratory reflectance measurements to routine field application was reported by Sanwald (39). The occurrence of rizomonia disease was investigated in sugar beet (Beta vulgaris) fields in West Germany. Laboratory reflectance measurements were made on leaves on "healthy" (uninfected) and diseased plants over the 0.5- and 2.5-

μm waveband. In some instances, an increase in reflectance of up to 100 percent was recorded for diseased plants, as compared with that for healthy plants, at the 0.55-, 0.59-, and 0.68- μm visible light wavelengths; differences in the near-infrared light reflectance were less than 5 percent, however. Sanwald found that the large increase in visible light reflectance was caused by the difference in canopy geometry between the diseased and healthy sugar beet plants. The leaves drooped on the diseased plants causing their whitish-appearing petioles to be exposed more to the remote sensor than were the petioles of healthy plant leaves. In essence, therefore, the visible light reflectance of petioles of diseased plant leaves was compared with the visible light reflectance of healthy plant leaves. This difference in reflectance was very vivid between the respective images on infrared color photographs. As a result, periodic remote sensing flights to detect rizomania-diseased sugar beets are now being made over West Germany and several other European countries (Personal Communique, Dr. Helmut Walter, University of Hohenheim, Stuttgart, West Germany).

Atmospheric Pollution

Most of the single-leaf spectral measurements have been conducted on simulated ozone-damaged plants in the laboratory. Thus, it should be kept in mind that there are many other atmospheric pollutants besides ozone and that ozone is found in a mixture of other pollutants in the "real world". An example of the results from laboratory spectrophotometric measurements made on single leaves is given below.

Ozone-damaged leaves develop necrotic areas which rapidly dehydrate. Thus, it was surmised that the detection of ozone-damage with remote sensing techniques would probably evaluate changes in leaf water content. This was substantiated by the following results.

Effects of ozone damage on the reflectance and photographic responses of cantaloupe (Cucumis melo) plant leaves and canopies were studied to determine the best wavelengths to detect ozone damage and to determine if ozone damage could be "previsually detected" (40). Mean spectrophotometrically measured light reflectances at the 0.55- and 0.65- μm wavelengths in the visible region (0.50- to 0.75- μm waveband) among the control and the lightly and severely ozone-damaged leaves was significantly greater ($P = 0.01$) than that for the other treatments. Mean reflectances among the treatments were not different statistically at the 0.85- μm wavelength in the near-infrared region (0.75- to 1.35- μm waveband).

The reflectances for the leaves of all treatments were different statistically ($P = 0.01$) for the 1.45-,

1.65-, 1.95-, and 2.2- μm wavelengths in the near-infrared water-absorption region (1.35- to 2.5 μm waveband). As ozone damage increased, reflectance increased because of leaf dehydration (41).

Field reflectance measurments over the 0.5- to 2.4- μm waveband for control and ozone-treated cantaloupe plant canopies supported the laboratory results. The reflectance of ozone-treated plants was the same as that for the control plants in the near-infrared plateau region (0.75- to 1.35- μm waveband) and in the near-infrared, water-absorption region (1.35- to 2.5- μm waveband). Therefore, the 1.65- and 2.2- μm wavelengths with atmospheric windows could be useful to detect ozone-damaged plants.

Practically, it might be possible to detect ozone-damaged plants in the water absorption region of the infrared spectrum (1.35- 2.5- μm waveband) because leaf dehydration greatly increases reflectance in this region.

Flourescence Detection

Flourescence measurement appears to be a promising tool to detect plant stress, particularly water stress (42). For example, decreased photosynthetic rates may be an indicator of stress conditions of plants. When photosynthetic rates are decreased with purportedly unchanging chlorophyll concentration, radiant energy absorbed by chlorophyll that is not used for carbon fixation is dissipated. Fluorescence is one mechanism that releases excess energy from chlorophyll.

Burke and Quisenberry (43) conducted an intensive investigation to determine if chlorophyll *a* fluorescence transients would provide a rapid, nondestructive technique to analyze the adaptation of photosynthetic light quanta conversion during severe water stress. The cotton strain "Lubbock Dwarf" was planted in an Amarillo loam soil which was irrigated to field capacity. A rainout shelter was used to keep water off the plots during the growing season, and selected rows were irrigated throughout the experiment with a drip-irrigation system. Measurements on chlorophyll *a* fluorescence, plant growth, plant water status, and photosynthesis were made on selected dates during the growing season.

The chlorophyll fluorescence rise was obtained using a portable fluorometer, model SF-20 (Richard Branker Research Ltd., Ottawa). The sensor of the fluorometer, which is placed directly on the leaf surface, contains both a light-emitting diode to irradiate the leaf surface (red light, 7 uE m^{-2} s^{-1}) and a photodiode to detect chlorophyll fluorescence emission.

The irrigated plants consistently had a lower fluorescence P/O ratio (P, maximum fluorescence level; O, initial fluorescence level) than that of the water-stressed plants. A typical P/O ratio for irrigated plants was about 1.3, while stressed plants routinely had a ratio of 1.6. A high P/O ratio shows that there is a better photosynthetic light quanta conversion, plants that exhibit a P/O of 1.0 are not photosynthetically active. The P-T/T ratio (P, maximum fluorescence level; T, steady state fluorescence level) was also utilized in analyses of the light quanta conversion, and the results supported the data provided by the P/O ratios. Common P-T/T ratios were 0.9 and 2.0 for irrigated and water-stressed plants, respectively.

The relationship between the fluorescence P/O ratio and the photosynthetic water-use-efficiency was also analyzed. A positive correlation between the P/O ratio and the water-use-efficiency of Lubbock Dwarf leaves was observed.

After three months of growth under water-stressed conditions, plants in the shelter were irrigated with 2-inches of water, and the P/O ratios were determined 48 hours after irrigation. Both irrigated and previously water-stressed plants had identical fluorescence and P/O ratios similar to those obtained for irrigated plants throughout the study.

They concluded that chlorophyll _a_ fluorescence transients can provide a rapid indication of the stressed-induced alteration of the photosynthetic light quanta conversion capabilities. The cotton genotype Lubbock Dwarf does undergo a modification of its photosynthetic system, which increases the efficiency of light quanta utilization under water-stressed conditions.

Estimating Constituents of Agricultural Products

Most agricultural applications using spectroscopy probably can be directly or indirectly traced back to single-leaf spectral measurements that were made in the laboratory (44, 45, 46, 47). A few applications of spectroscopy are indicated below.

The use of near-infrared reflectance is a promising tool to measure various constituents in plant materials and agricultural products. Three selected examples are: Estimating nutrient content of dehydrated vegetables (48); analysis of nitrogen and oil content of plant materials (49), and analysis of forage crop quality (50).

Near-infrared analysis is based on the utilization of energy absorbed in the near-infrared region of the electromagnetic spectrum by chemical groups that are characteristic of major constituents of the plant material or agricultural product. After proper calibration of the instrument using samples with known concentration of the desired component to be measured, interference by other constituents is kept minimal by using multiple wavelength readings, which are incorporated or integrated into multiple regression equations for prediction purposes.

Further research is still underway in this field. In particular, studies are being conducted on fieldgrown crops to see if plant leaf spectral properties can be used to predict plant production of, for example, protein. This potential application may be useful in plant-breeding programs.

Plant Breeding Programs

Several methods have been proposed as potential valuable tools to assist plant breeders in developing stress-resistant crops by screening for characters and/or considering factors such as : nitrogen, protein, and oil content in segregating populations of crop plants (51, 52, 53); tolerance to water stress conditions (54, 55); stress indices to predict yields with infrared thermometry (56); biomass production (57, 58); and improved evaluation of plant-stress tolerance by measuring leaf expansion (59).

Research has shown that leaf spectral measurements should be useful to screen cotton strains for leaf chlorophyll concentration, water content, and structure, providing that leaves are about the same age. For example, as leaf chlorophyll concentration increases, reflectance decreases at the 0.55- μm wavelength (60, 61); leaves with a high water content have less reflectance at the 1.65- μm wavelength than leaves with a lower amount of water (62); and leaves with a lacunose mesophyll (many air spaces) have more near-infrared reflectance (0.75- to 1.35- μm waveband) than do more compact leaves (63). For screening to be feasible, however, care must be taken so that the spectral measurements are made on leaves of the same chronological age.

The potential usefulness of leaf spectral measurements to screen cotton (*Gossypium hirsutum*) strains for characters affected by stress was investigated (64). Laboratory spectral measurements [reflectance, transmittance, and absorptance] were made on upper (adaxial) leaf surfaces of irrigated and photoperiodic cotton strains; 'T25', drought tolerant and 'T169', drought susceptible; that were grown in a rainout shelter in Lubbock, Texas in 1982.

Leaves of irrigated T25 plants had lower reflectance at the 0.55- μm wavelength than did T169 leaves of irrigated plants, which appeared to be related to their total chlorophyll concentrations. The T25 leaves from irrigated plants had more reflectance at the 0.85- μm wavelength than did T169 leaves of nonirrigated plants. Apparently, T25 leaves were more lacunose and succulent than T169 leaves. In the water absorptance region, 1.35- to 2.5- μm waveband, the thinner leaves of the T169 strain had higher transmittance values, for example, than did the thicker leaves of the T25 strain. These results indicated that screening of cotton germplasm with spectral measurements for characters such as leaf chlorophyll concentration, thickness, and water content should be feasible by using spectral measurements.

Detection of Herbicidal Effects

The detection of herbicidal effects on plants is a relatively new and promising use of single-leaf reflectance measurements. Some of the pioneering research was conducted by Walter and Koch (65). They reported the effects of atrazine (1.44 liter/ha rate) and 2,4-D amine (.72 liter/ha rate) on beans (*Phaseolus vulgaris*) and maize (*Zea mays*).

Beans but not maize were very sensitive to both herbicides. Two days after treatment the effects of both herbicides on leaf light reflectance were detectable spectrophotometrically, even though there were no noticeable visual differences between treated and untreated leaves. Apparently, this is another example of promising "previsual detection".

Eight days after herbicidal application, there were large spectrophotometrically measured differences among the treated and untreated leaves. At this time, marked visual differences were evident too. At the 0.55- μm wavelength (green light reflectance peak), reflectance of treated leaves was 50 and 46 percent higher than that of the control leaves for the 2,4-D (2,4-dichlorophenoxyacetic acid) and atrazine treatments, respectively. The near-infrared reflectance at the 0.80- μm wavelength, which is strongly affected by mesophyll structure, was 12 percent lower for the 2,4-D treated leaves and 7 percent lower for the atrazine-treated leaves, as compared with the reflectance of the control leaves. At the 2.2- μm wavelength in the infrared light water absorption region, atrazine-treated leaves had 35 percent higher reflectance than did the untreated leaves. This result was apparently caused by differences in leaf water content: atrazine-treated and untreated leaves had 67 and 85 percent water content on an oven-dried weight basis, respectively. High infrared light reflectance in the water absorption region is associated with low leaf water content; whereas, lower reflectance is associated with higher leaf water content.

More research must be conducted, however, to field-test the laboratory measurements on the detection of herbicidal effects on plants.

Carbon Dioxide Concentration

Aircraft-mounted sensors were used to measure carbon dioxide exchange above a cornfield, forest, and lake during midday conditions (66). This sensor has the adaptability to make carbon dioxide measurements on individual leaves. Mean carbon dioxide absorption values were consistent with groundbased observations. Therefore, it was speculated that such information could be used to quantitatively evaluate carbon dioxide source and sink distributions in the biosphere, to correlate satellite with near-surface measurement data, and to monitor crop performance such as phytomass production. It is also probable, that the detection of a difference in oxygen concentration might be a very early indicator of the onset of a plant stress condition. This would have very important implications in management of agricultural resources.

Photoacoustic Spectroscopy

A relatively new technique, photoacoustic spectroscopy, has been adapted to the measurement of plant leaf absorption (67). These investigators reported that the intensity of anthocyanin could be characterized in leaves of three *Euphorbia* species by using photoacoustic spectroscopy. With this instrumentation, a sample without prior preparation is placed in a gas-filled, closed chamber containing a microphone. After the sample is illiminated by monochromatic light, a part of the radiation absorbed by the sample is converted into heat, which causes pressure waves and transmits the signals to a recorder.

Interestingly, the investigators postulated that anthocyanin pigmentation in leaves may be an adaptation of a plant to water stress. The degree of intensification was correlated inversely with the soil and plant water content and leaf water potentials. Accordingly, photoacoustic spectroscopy might have applicability to water-stress detection in plants.

Plant Growth Regulators

Possibly, plant growth regulators should not be treated in a separate section. However, their versatility to modify plant's growth and stress tolerance is becoming widely recognized. In the World's Eastern Hemisphere, for example, about 5 million hectares of small grain were treated with Cycocel in 1981 (Personal Communication, Mr. Kienreich, West Germany, 2 December 1981). Cycocel increases small grain yields by decreasing lodging, particularly when high fertilizer rates are used (68, 69). Therefore, the plant growth regulator effects on plant growth and stress provides the potential for a remarkable "research tool" to further understand remote sensing. Two examples follow.

Mepiquat chloride (1,1-dimethylpiperidinum chloride) caused treated cotton (*Gossypium hirsutum*) leaves to become thicker than untreated leaves, which increased near-infrared light reflectance (70). Also, mepiquat chloride-treated leaves had a higher chlorophyll concentration than untreated leaves. Subsequently, visible light reflectance was decreased. This result led to more basic studies. For example, it was first speculated that mepiquat chloride increased the total leaf chlorophyll concentration because it also increased leaf thickness. When this premise was found not to be true, studies were conducted on leaf chloroplasts. These results (unpublished) indicated that cells of mepiquat chloride-treated leaves had a greater number of chloroplasts with more starch grains in each chloroplast than did the cells of nontreated leaves. The above is only an example that shows the possible value of spectrophotometrically measured leaf reflectance to assess biological activity of chemicals. Moreover, near-infrared light reflectance measurements are useful too. For example, a bioregulator that reduced leaf expansion might be detectable, because a compact leaf usually has less reflectance than that of a "fully-expanded" leaf. In our work, for example, mepiquat chloride enhanced the tolerance of undesirable low and high temperatures (71, 72). This was easily detectable by reflectance measurements and led us to study effects of the chemical on the plasma membrane.

Mepiquat chloride was used to compare the reflectance of treated (reduced growth) with that of nontreated ("normal") cotton plant canopies (73). Radiometric canopy and laboratory spectrophotometric single-leaf reflectance measurements were used to compute the reflectance differences, at four sample dates through the 1980 growing season, between nontreated and mepiquat chloride-treated cotton. The single-leaf reflectance measurements for the mepiquat chloride-treated cotton plants were lower at the 0.65- μm wavelength and higher at the 0.85-μm wavelength than the nontreated cotton plants for three of the four sample dates. Early in the growing season the canopy reflectance of mepiquat chloride-treated cotton was lower in the red and higher in the near-infrared reflectance of nontreated cotton canopies. Later in the season the canopy reflectance measurements indicated that the nontreated cotton plants senesced faster than the mepiquat chloride-treated cotton plants (74). These results should provide a basic understanding of temporal canopy reflectance obtained from untreated and mepiquat chloride-treated cotton for use by operational remote sensing.

Conclusion

Single-leaf spectral measurements in the laboratory have contributed greatly to an understanding of the mechanism of the interaction of electromagnetic radiation with plant leaves and other plant components such as leaf petioles, leaf sheaths, heads of grasses, inflorescences, *etc*. This has included spectrophotometric measurements of leaf reflectance, transmittance, and absorptance, generally over the 0.5- to 2.5- μm (sometimes 0.4- to 2.6- μm waveband) waveband.

The laboratory spectrophotometric measurements have been used to characterize visible light spectra (0.45- to 0.75- μm waveband), near-infrared light spectra (0.75- to 1.35- μm waveband), and infrared water absorption light spectra (1.35- to 2.5- μm waveband) for biological factors such as leaf pigment concentration, pubescence, and senescence. Probably, the best indicator of changes in leaf optical properties was a difference in reflectance at the 0.55- μm wavelength caused by variations in leaf pigment concentration: High reflectance was associated with low total chlorophyll concentration; and conversely, low reflectance was correlated with high chlorophyll concentration.

Near-infrared light is scattered or reflected from leaves by refractive index discontinuities. The most important discontinuity is the cell wall/air-space interface. If near-infrared light travels at a critical angle from a hydrated cell wall with a refractive index of about 1.4 to an open air space with a refractive index of 1.0, the near-infrared light is scattered or reflected. This can be easily demonstrated by replacing air in leaves with a liquid by vacuum infiltration. Reflectance becomes minimal if the refractive index of the liquid presumably matches the refractive index of the cell wall.

It is extremely important to consider leaf maturity if spectrophotometric measurements are to be made to compare the spectral properties of leaves. Young leaves have a more compact structure than nonstressed leaves; stressed leaves are not fully expanded and have a more compact structure than fully expanded, nonstressed leaves.

Generally, stressed plant leaves have a lower near-infrared light reflectance than do nonstressed leaves. However, spectrophotometric measurements have shown that diseased leaves may have higher near-infrared light reflectance than "normal" leaves. This is probably caused by leaf dehydration, and the damaged tissue collapses as such a manner that the number of air voids in the leaf mesophyll increases and light scattering correspondingly increases.

Internal discoloration of leaves and/or black coating on their surface will cause a decrease in near-infrared light reflectance. The discoloration can be caused by saponification of chlorophyll and the oxidation and polymerization of polyphenol oxidase to a brown pigmentation. This change in reflectance is not caused by a change in internal air spaces, but the absorptance of near-infrared light is increased by internal chemical changes or by chemicals in the black coatings on leaf surfaces.

Single-leaf measurements have been used successfully

to a very limited extent in the detection of stresses such as low and high temperature, soil salinity, nutrient deficiency or toxicity, plant maladies, and atmospheric pollution. However, the single-leaf measurements have been used rather extensively to study irrigation scheduling, plant-canopy interactions, phytomass, and constituents of agricultural products. This technique may also aid in the selection of desirable plant traits in plant breeding programs.

There are new and promising stress detection techniques; namely, fluorescence detection, measurement of carbon-dioxide concentration, and photoacoustic spectroscopy.

Bioregulators are known to affect leaf pigmentation and expansion. Thus, spectrophotometric measurements on leaves from treated plants might indicate activities of chemicals and give a clue to their effects on bioregulation.

With the availability of more sophisticated instrumentation, it seems feasible that more subtle differences at various wavelengths of light reflectance will be detectable among plant leaves.

Literature Cited

1. Turner, N. C.; Kramer, P. J. John Wiley & Sons, New York, New York, 1980. 482 pp.
2. Gausman, H. W. Photogram. Eng. 40:183-192. 1974.
3. Nixon, P. R.; Namken, L. N.; Wiegand, C. L. Remote Sensing Earth Resources 2:643-657. 1973.
4. Pinter, P. J., Jr.; Jackson, R. D.; Idso, S. B.; Reginato, R. J. Int. J. Remote Sensing 2:43-48. 1981.
5. Walter, H.; Gausman, H. W.; Escobar, D. E.; Rodriguez, R. R.; Rittig, F. R. J. Rio Grande Valley Hort. Soc. 35:27-33. 1982.
6. Escobar, D. E.; Bowen, R. L.; Gausman, H. W.; Cooper, G. R. J. Rio Grande Valley Hort. Soc. 36:61-66. 1983.
7. Peynado, A.; Gausman, H. W.; Escobar, D. E.; Rodriguez, R. R.; Garza, M. V. Cryobiology 16:63-68. 1979.
8. Hoffer, R. M.; Johannsen, C. J. Remote Sensing in Ecology P. O. Johnson, ed., Univ. of Georgia Press, Athens, Georgia, pp. 1-16. 1969.
9. Thomas, J. R.; Namken, L. N.; Oerther, G. F.; Brown, R. G. Agron. J. 63:845-847. 1971.
10. Namken, L. N. Agron. J. 57:38-41. 1965.
11. Thomas, J. R.; Namken, L. N.; Oerther, G. F.; Brown, R. G. Agron. J. 63:845-847. 1971.
12. Carlson, R. E.; Yarger, D. N.; Shaw, R. H. Agron. J. 63:486:489. 1971.
13. Lesage, P. Rev. Gen. Botan. pp. 168-170. 1908.

14. Harter, L. L. U.S. Dept. of Agr. Bur. Plant Ind. Bul. 134. 1908.
15. Chermezon, H. Ann. Sci. Nat. Bot. 12:117-129. 1910.
16. Hayward, H. W.; Long, E. M. Bot. Gaz. 102:437-462. 1941.
17. Hayward, H. W.; Bernstein, L. Botan. Rev. 24:584-635. 1958.
18. Strogonov, B. P. Physiological Basis of Salt Tolerance of Plants; 1962. Daniel Davey and Co., Inc., New York, New York. Translation by Israel Program for Scientific Translations, 1964. 279 pp.
19. Nieman, R. H. Plant Physiol. 40:156-161. 1965.
20. Gausman, H. W. Photogram. Eng. 40:183-192. 1974.
21. Fritz, N. L. Photogram. Eng. 33:1128-1138. 1967.
22. Nixon, P. R.; Lyda, S. D.; Heilman, M. D.; Bowen, R. L. Texas Agric. Exp. Sta. Misc. Pub. 1241, 1975. 4 pp.
23. Thomas, J. R.; Oerther, G. F. Agron. J. 64:11-13. 1972.
24. Gausman, H. W.; Gerbermann, A. H.; Wiegand, C. L. Photogram. Eng. & Remote Sensing 41:177-181. 1975.
25. Gausman, H. W.; Escobar, D. E.; Rodriguez, R. R. Proc. 4th Biennial Workshop Color Aerial Photogr. pp. 13-27. 1973.
26. Escobar, D. E.; Gausman, H. W. J. Rio Grande Valley Hort. Soc. 32:81-88. 1978.
27. Colwell, R. N. Hilgardia 26:223-286. 1956.
28. Keegan, H. J.; Schleter, J. C.; Hall, W. A., Jr.; Hass, Gladys M. Nat'l. Bur. Stds. Rpt. 4591. 1956. 161 pp.
29. Gausman, H. W.; Cardenas, R. Bot. Gaz. 130:158-162. 1969.
30. Manzer, F. E.; Cooper, G. R. Maine Agric. Exp. Station Bull. 646:1-4. 1967.
31. Burns, E. E.; Starzyk, M. J.; Lynch, D. L. Trans. Illinois State Acad. Sci., V. 62. 1969.
32. Gausman, H. W.; Escobar, D. E.; Rodriguez, R. R.; Thomas, C. L.; Bowen, R. L. Photogram. Eng. & Remote Sensing 44:481-485. 1978.
33. Heller, R. C. Proc. 5th Symp. Remote Sens. Environ., Univ. of Michigan, Ann Arbor, pp. 387-434. 1968.
34. Meyer, Merle P. Proc. Workshop Infrared Color Photography. Part V, pp. 5-7. 1967.
35. Cardenas, R.; H. W. Gausman; W. A. Allen; Marcia Shupp. Remote Sensing Environ. 1:199-202. 1969-1970.
36. Keegan, H. J.; J. C. Schleter; W. A. Hall, Jr.; Gladys M. Hass. Nat'l. Bur. Stds. Rpt. 4591. 1956. 161 pp.
37. Gausman, H. W.; R. Cardenas. Bot. Gaz. 130:158-162. 1969.

38. Escobar, D. E.; R. L. Bowen; H. W. Gausman. Remote Sensing Environ. 1982.
39. Sanwald, E. F. In: Spectral Signatures of Objects in Remote Sensing, International Colloquium, Avignon, France. pp. 201-208. 1981.
40. Gausman, H. W.; D. E. Escobar; R. R. Rodriguez; C. L. Thomas; R. L. Bowen. Photogram. Eng. & Remote Sensing 44:481-485. 1978.
41. Gausman, H. W. Photogram. Eng. 40:183-192. 1974.
42. McFarlane, J. C.; R. D. Watson; A. F. Theisen; R. D. Jackson; W. L. Ehler; P. J. Pinter, Jr.; S. B. Idso; R. J. Reginato. Appl. Opt. 19:3287-3289. 1980.
43. Burke, J. J.; J. E. Quisenberry. Beltwide Cotton Prod. Res. Conf. Prod., pp. 60-61. 1983.
44. Seybold, A. Planta 16:195-226. 1931.
45. Seybold, A. Planta 18:479-508. 1932.
46. Gates, D. M.; H. J. Keegan, J. C. Schleter; V. R. Weidner. Appl. Opt. 4:11-20. 1965.
47. Keegan, H. J.; J. C. Schleter; W. A. Hall, Jr.; Gladys M. Hass. Nat'l. Bur. Stds. Rpt. 4591. 1956. 161 pp.
48. Park, Y. W.; M. J. Anderson; A. W. Mahoney. J. Food Sci. 47:1558-1561. 1982.
49. Starr, Carol; A. G. Morgan; D. B. Smith. J. Agric. Sci. Camb. 97:107-118. 1981.
50. Norris, K. H.; R. F. Barnes; J. E. Moore; J. S. Shenk. J. Animal Science 43:889-897. 1976.
51. Norris, K. H.; R. F. Barnes; J. E. Moore; J. S. Shenk. J. Animal Science 43:889-897. 1976.
52. Starr, Carol; A. G. Morgan; D. B. Smith. J. Agric. Sci. Camb. 97:107-118. 1981.
53. William, P. C.; S. G. Stevenson; P. M. Starkey; G. C. Hawtin. J. Sci. Fd. Agric. 29:285-292. 1978.
54. Blum, A.; F. Schertz; R. W. Toler; R. I. Welch; D. T. Rosenow; J. W. Johnson; L. E. Clark. Agron. J. 70:472-477. 1978.
55. Blum, A.; J. Mayer; G. Gonzlan. Field Crops Res. 2:137-146. 1982.
56. Idso, S. B.; R. J. Reginato; R. D. Jackson; P. J. Pinter, Jr. Irrig. Sci. 2:205-212. 1981.
57. Tucker, C. J.; L. D. Miller; R. L. Pearson. Remote Sens. Earth Res. 2:601-626. 1973.
58. Tucker, C. J.; L. D. Miller; R. L. Pearson. Photogram. Eng. & Remote Sensing 41:1157-1162. 1975.
59. Karamanos, A. J.; J. Elston; R. M. Wadsworth. Ann. Bot. 49:815-826. 1982.
60. Rabideau. G. S.; C. S. French; A. S. Holt. Am. J. Bot. 33:769-777. 1946.
61. Thomas, J. R.; H. W. Gausman. Agron. J. 69:799-802. 1977.

62. Gausman, H. W.; E. D. Knipling; D. E. Escobar. Photogram. Eng. & Remote Sensing. 43:1183-1185. 1977.
63. Gausman, H. W.; W. A. Allen; R. Cardenas; A. J. Richardson. Appl. Opt. 9:545-552. 1970.
64. Gausman, H. W.; J. E. Quisenberry; J. J. Burke; C. W. Wendt. Field Crops Res. 1984. (In press).
65. Walter, H.; W. Koch. Proc. British Crop Protection Conference--Weeds. 1980. 8 pp.
66. Brach, E. J.; R. L. Eesjardins; P. Alvo; P. H. Schuepp. Science 216:773-735. 1982.
67. Veeranjaneyulu, K.; V. S. R. Das. J. Exp. Bot. 33:515-519. 1982.
68. Jung, J. U.; F. Sturn. Landwirtsch. Forsch. 17:1-9. 1964.
69. Mathews, P. R.; J. J. B. Caldicott. Ann. Appl. Bot. 97:227-236. 1980.
70. Gausman, H. W.; L. N. Namken; E. Stein; R. W. Leamer; H. Walter; R. R. Rodriguez; D. E. Escobar. J. Rio Grande Valley Hort. Soc. 33:105-112. 1979.
71. Richardson, A. J.; H. W. Gausman. Remote Sensing Environ. 12:501-507. 1982.
72. Gausman, H. W.; D. E. Escobar; R. R. Rodriguez. Plant Growth Regulator Soc. Am. Bull. 9:6-8. 1981.

RECEIVED February 22, 1984

Chemical Modification of Plant Response to Temperature Extremes

MERYL N. CHRISTIANSEN and JUDITH B. ST. JOHN

Agricultural Research Service, U.S. Department of Agriculture, Beltsville, MD 20705

Lipid synthesis in vegetative tissue can be modified by pyridazinones. The major effect is blockage of desaturation of linoleic acid to linolenic acid. The blockage of linolenic acid synthesis inhibits cold hardening and increases chilling sensitivity of cotton and freezing susceptibility of winter grains. Other modifications of lipid synthesis include an increase in leaf cuticular wax up to 55%. Foliar wax increase improves water economy of soybean, corn, sorghum and cotton.

Temperature is the primary environmental factor that limits where and how successfully agricultural crops can be grown. There is a wide range of genetic diversity among economic crop species in ability to tolerate temperature extremes. Many tropical species are injured by temperatures above 0°C, while temperate zone species can withstand temperatures as low as -70°C. An understanding of the nature of low temperature resistance or of chilling and freezing injury is essentially lacking.

Various studies have associated membrane form and function with low temperature relations. A variety of physiological dysfunctions of chilled and frozen plants have been related to membrane perturbation (1). Many of the changes induced by low temperature are characterized by changes in lipids. The changes may be an alteration of classes such as an increase in total polar lipid, or of triglycerides or only an alteration of constituent fatty acids. A declining temperature is almost universally effective in increasing fatty acid unsaturation. Reduction in oleate and linoleate and an increase in linolenate is usual in winter hardening plants (2).

An increase in functionality of cells at low temperature may be concurrent with increased fatty acid unsaturation . Processes such as water transport, respiration, and photosynthesis in hardened tissue are less inhibited by low temperatures. The function of membranes is requisite for osmotic control of cell contents and of water. Resistance to chilling, frost, and freezing is dependent upon a membrane-regulated orderly dehydration of cells with a concurrent rise in cell osmotic concentration.

Chemical regulation of fatty acid should, therefore, provide a tool to manipulate plant response to temperature. Such an ability to regulate plant response to temperature would obviously be useful to abate cold and heat damage to crop plants. It would likewise be useful to control unwanted vegetation (weeds) by inducing greater susceptibility to heat or cold.

There is little past history concerning the use of chemicals to alter lipids in plants. Alteration of fatty acid composition of micro-organisms by selective fatty acid supplements has provided evidence that temperature response can be altered by changes in lipid components. McElhaney (3) showed that treatment of Aspergillis laidlawii with fatty acid supplements shifted the membrane phase transition to a different temperature level. Membrane lipid alteration as a result of temperature acclimation is common in plants, animals and microorganisms and has been related to fluidity of membranes at low temperature in Tetrahymena pyriformis (4,5). In these studies the activities of membrane bound fatty acid desaturases were controlled by membrane fluidity at low temperature. Fatty acid composition of membrane phospholipids can reflect the lipid composition of the culture or diet; e.g. animals fed corn oil (6), yeast grown in media of diverse fatty acids (1); or potato tuber tissue incubated in media of varying fatty acid unsaturation or chain length (7). Treatment of tomato (Lycopersicum esculentium) with ethanolamine-tween-oleate altered phospholipid composition and reduced chilling injury symptoms (8).

Considerable controversy has prevailed over the interrelation of membrane fatty acid unsaturation, membrane fluidity at low temperature and plant resistance to chilling or freezing (1). Some researchers (2) hold that there is no relation between lipid unsaturation and plant survival at low temperature. They cite as evidence the lack of differences in lipid unsaturation between freeze hardy and unhardy wheat varieties. Other researchers suggest a universal need for unsaturated lipids in membranes to insure fluidity and hence functionality at low temperature. Unless fluidity and consequent functionality exists, subsequent cold hardening events cannot occur (9).

Pyridazinone compounds were first reported to alter production of polar lipids of chloroplasts (10). In subsequent work, St. John (11) identified pyridazinones that specifically inhibited production in vegetative tissue of linolenic acid with concurrent increases in linoleic acid. The pyridazinones appeared to be excellent tools to settle the dispute over the significance of fatty acid unsaturation in the cold hardening process.

The initial experiments were concerned with chilling injury to seedling cotton. Chilling injury in cotton seedlings is caused by cold inactivation of root water uptake; the plant suffers dehydration (12). Cotton seedlings can be chill hardened to tolerate 5-8°C by gradually lowering the temperature from 20° to 8°. The questions we posed were: (1) Does linolenic acid increase in cotton roots with reduction in temperature? (2) Will the pyridazinones block a low temperature induced increase in linolenic acid? (3) If linolenic acid synthesis is blocked will the response to low temperatures be altered in terms of sensitivity to chilling and ability to harden?

Seeds were germinated at temperatures of 15, 20, 25 and 30° in paper rolls wet with either water or BASF 13338, [4-chloro-5(dimethyamino-2 phenyl-3 (2H)-pyridazinone] at 10 um concentration. Seeds were germinated 24 hours at 30°, 40 hours at 25, 70 hours at 20° and 168 hours at 15°C to insure equally sized seedlings. Root tip (1 cm) and hypocotyl (2 cm) were collected and freeze dried for lipid analysis as previously described (13). Effects of temperature and BASF 13338 treatment on seedling growth were determined on seeds germinated at 30°, cultured in growth chambers and treated with water or 20 uM BASF 13338 in the culture media [5 ug/g (w/w)]. Seedlings were chill hardened by gradually lowering temperature (5°/48 hr) to 15°. A control and BASF 13338 treatment were not hardened but grown 8 days at 30°. After the above temperature sequences, all treatments were chilled 4 days at 8°, then grown for seven days in the greenhouse to determine survival and injury symptoms.

The fatty acids of the polar lipids of the root tips were progressively higher in linolenic acid as growth temperature was reduced (Table I). BASF 13338 essentially blocked low temperature induction of linolenic acid synthesis and caused a concurrent increase in linoleic acid. Seedlings with reduced levels of membrane-bound linolenic acid were less able to survive chilling temperature and surviving seedlings were abnormal (Table II) (13).

Table I. Fatty Acid Composition of Polar Lipids of 1-cm Root Tips of Cotton Seedling Radicles (13)

Treatment	Growth Temperature °C	Fatty Acid (%) 16:0	18:0	18:1	18:2	18:3
Control (water)	30	32.6	3.7	7.5	36.3	13.6
	25	35.3	2.8	5.7	33.4	22.9
	20	32.9	3.3	5.4	31.7	26.8
	15	31.6	4.8	4.2	31.9	27.5
BASF 13338 (10 uM)	30	34.7	2.8	7.6	44.0	10.9
	25	34.4	2.7	5.7	43.1	14.0
	20	34.5	3.7	6.4	41.7	13.7
	15	30.9	4.1	4.6	43.7	17.1

Table II. Cotton Seedling Survival 7 Days after 96 Hr of Chilling at 8°C (13)

Growth Conditions	Seedling Classification Normal	Abnormal[1]	Dead
		%	
Control 30°C, 8 days	62	35	3
BASF 13338 (10 uM) 30°C, 8 days	0	60	40
Control, hardened[2]	92	9	0
BASF 13338 (10 uM), hardened	80	20	0

1 Seedlings with 50% of cotyledon necrotic were classed abnormal.
2 Grown 2 days at 30°C, 2 days at 25°C, 2 days at 20°C, and 2 days at 15°C.

After ascertaining an association between linolenic acid and chilling sensitivity in cotton, research to determine the essentiality of linolenate to freeze hardening in winter grains was initiated. Willemot (14) reported that both the accumulation of linolenic acid and development of freezing tolerance were inhibited when wheat seedlings were treated with BASF 13338 thirty-six hours before frost hardening. These findings were extended into a field situation (15). BASF 13338 was soil incorporated at seedling time at rates of 0, 2.8, 5.2 and 11.2 kg/ha. Wheat (Triticum aestivum L. cvs. Arthur and Potomac), barley Hordeum vulgare L., and rye (Secale cereale L.) were sown in the fall. Visual evidence of cold injury of BASF 13338 treated plants became evident after the first frost. Severity of cold injury increased with increased rate of BASF 13338 application.

Fatty acid composition of membrane lipids from seedling shoots of BASF 13338 treated plants had significantly lower levels of linolenic acid (Table III). Reduced level of linolenic acid was directly related to reduced plant survival (Table IV) and reduced tillering of surviving plants (Table V). There was evidence of varietal and genus differences in sensitivity to BASF 13338. The decreasing order of sensitivity was Authur, Potomac wheats, Monroe barley and Abruzzi rye. The data indicate a relation between linolenic acid and winter survival of cereals. Other evidence suggests that in winter cereals, linolenic acid is not the limiting factor in the minimal temperature tolerated.
De La Roche (2) has suggested that it is the inhibition of photosynthesis by pyridazinones (10) that is responsible for increased cold sensitivity. However, other pyridazinones inhibit photosynthesis (Pyrazon); but do not affect linolenic acid or reduce cold hardening of winter wheat. Also, pyridazinones reduce linolenic acid levels in non-photosynthetic tissues (13).

Table III. Effect of Preemergence Treatment with BASF 13338 on Membrane Fatty Acid Composition of Shoot Tissue from Small Grains (15)

Species	BASF 13338	Fatty acid composition by weight[1] (%)				
		16:0	18:0	18:1	18:2	18:3
	kg/ha					
Arthur wheat	0.0	10.1 a	0.8 a	1.8 a	11.7 a	75.6 a
	2.8	8.8 b	0.8 a	2.4 a	35.4 b	52.8 b
	5.6	8.4 b	0.6 a	2.1 a	50.0 c	38.9 c
	11.2	8.7 b	0.8 a	2.2 a	61.4 d	27.1 d
Potomac wheat	0.0	10.0 a	1.1 a	1.9 a	11.7 a	75.4 a
	2.8	9.1 a	1.0 a	2.6 a	35.4 b	52.4 b
	5.6	8.8 a	1.0 a	2.3 a	44.0 c	44.0 c
	11.2	9.4 a	1.2 a	2.4 a	54.5 d	32.5 d
Monroe barley	0.0	11.6 a	1.3 a	1.7 a	10.0 a	75.4 a
	2.8	10.0 b	1.0 a	1.4 a	20.8 b	66.9 b
	5.6	10.4 ab	1.2 a	1.7 a	31.9 c	54.9 c
	11.2	10.4 ab	1.3 a	1.5 a	40.2 d	46.6 d
Abruzzi rye	0.0	11.3 a	1.0 a	2.4 a	14.2 a	71.3 a
	2.8	10.2 ab	0.8 a	2.1 a	18.4 b	65.7 b
	5.6	9.8 bc	0.9 a	2.4 a	32.4 c	54.5 c
	11.2	8.8 c	0.9 a	1.7 a	30.8 c	57.6 c

[1] Values for individual fatty acids within a cereal followed by common letters are not significantly different at the 5% level with Duncan's Multiple Range Test.

Table IV. Effect of BASF 13338 on the survival of winter cereals following a freeze at -5°C (15)

Species	Survivial (%)	
	Control	BASF 13338 (10 ug/g)
Arthur wheat	96	29
Potomac wheat	100	0
Monroe barley	93	0
Abruzzi rye	93	0

Table V. Effect of BASF 13338 Applied Preemergence on Tillering of Winter Cereals During Winter of 1977-78 (15)

Species	BASF 13338 (kg/ha)			
	0	2.8	5.6	11.2
	---------Mean no. of tillers/m of row-------			
Arthur wheat	224 a[1]	174 b	124 c	11 d
Potomac wheat	183 a	173 a	60 b	5 c
Monroe barley	164 a	138 a	60 b	3 c
Abruzzi rye	199 a	133 b	99 b	9 c

[1] Values followed by common letters are not significantly different at the 5% level with Duncan's Multiple Range Test.

Treatment with BASF 13338 reduced the proportions of linolenic acid in phosphatidylcholine (Table VI) phosphatidylathanolamine, and the total polar lipids of cereal roots at all growth temperatures (16). The distribution of phospholipid classes was not influenced by temperature or treatment with BASF 13338 (Table VII). Proportions of linolenic acid in root membranes were not related to the rate of respiration from 4 to 30C, the activation energies of respiration, or the temperature at which the change in activation occurred. Proportions of linolenic acid in root membranes of cereals may limit other membrane-associated processes such as active transport, water and ion permeability, and membrane resilience.

These collective studies indicated that the pyridazinones alter the response of cotton and cereals to low temperatures. We conjectured that the pyridazinones might also alter the response of some plant species to high temperatures. Field tests with Silver Queen corn indicated that preemergence treatment with BASF 13338 ameliorated high temperature stress, probably by alternation of water status (17). Growth chamber studies were conducted on the heat tolerance of corn (Zea mays L.) and sorghum [Sorghum bicolor (L.) Moench]. Corn and sorghum plants were grown at 30°C and transferred to 40°C. Plants treated with BASF 13338 transpired significantly (P = 0.01) less water than untreated controls at 40°C; the measured rates were 0.59 g H_2O; cm^2 leaf area/24 hr and 0.72 g H_2O/cm^2 leaf area/24 hr, respectively. Soybean and cotton plants treated with BASF 13338 transpired 20-30% less water than control plants (Table VIII). The difference in water loss did not seem to involve stomates since there were no differences in diffusive resistance or pressure bomb readings between control and BASF 13338 treated leaves. Furthermore, internal leave water content, determined gravimetrically, did not differ between leaves of control and treated plants. Pyridazinone-induced increases in cuticular waxes may explain the alterations in water status. The cuticular wax of soybeans grown in the greenhouse in nutrient solutions containing 15 ppm BASF 13338 were increased 40% over control levels. Pyridazinone treatment increased cuticular waxes of soybean leaves by as much as 55% under field conditions (Table IX). Thus the evidence seems to indicate that BASF 13338 does alter the water economy of plants, probably through effects on wax deposition on leaves and cuticular water relations.

Table VI. Fatty Acid Composition of Phosphatidyl Choline from Roots of Wheat Seedlings (16)

Treatment	Growth Temperature	Fatty Acid				
		16:0	18:0	18:1	18:2	18:3
	°C	% by weight				
Control	25	27	2	5	49	16
	20	22	3	10	44	21
	15	26	5	9	34	26
	10	24	2	6	32	36
BASF 13338 (100 uM)	25	25	2	5	63	5
	20	23	3	7	61	5
	15	21	4	9	58	7
	10	19	2	5	63	11

Phospholipids were separated using high performance liquid chromatography. The phosphatidyl choline fraction was collected and the fatty acid composition determined.

Chemical modification of plant temperature relations are possible through alteration of polar lipids of membranes. Use of pyridazinone to inhibit fatty acid desaturases results in lower levels of linolenic acid in polar lipids and a concurrent reduction of plant tolerance to chilling in tropical species and freezing in winter grain crops. Alteration in lipids also confers tolerance to "summer stress" and improved water economy which can be attributed to increases in leaf cuticular wax which reduces cuticular transpiration.

Table VII. Phospholipid Composition of Wheat Seedling Roots (16)

Treatment	Growth Temperature °C	Phospholipid Fractions[1]				
		PE[2]	PG	PI	PA-PS	PC
		% lipid phosporus				
Control	25	34	5	7	4	51
	20	37	5	4	6	48
	15	31	8	7	3	52
	10	38	6	7	4	45
BASF 13338 (100 uM)	25	35	6	5	4	50
	20	33	6	5	2	54
	15	34	5	7	4	50
	10	37	5	4	3	51

1 Phospholipids were separated into five fractions using high performance liquid chromatography. Fractions were collected and the phosphorus content determined by Bartlett's modification of the Fiske-Subbarow method (18)

2 PE = phosphatidylethanstamine, PG = phosphatidylglycerol; PI = phosphatidylinositol; PA-PS = phosphatidic acid + phosphatidylserine; PC = phosphatidylserine.

Table VIII. Effect of BASF 13338 on Water Less by Cotton and Soybean at 30°C

Treatment	Water Use % of Control
15 ppm BASF 13338 in Nutrient Solution (Cotton)	72
15 ppm BASF 13338 in Nutrient Solution (Soybean)	78
Soil - BASF 13338 - 50 mg/kg (Soybean)	76

Table IX. Effect of BASF 13338 on Cuticular Wax of Field Cultured Clark Soybean

BASF 13338 Rate kg/ha	Leaf Wax ug/cm^2 Leaf	% Increases (%)
0	4.79	
5	5.29	12
10	7.43	55

Literature Cited

1. Lyons, J. M. Ann. Rev. Plant Physiol. 1973, 63, 445-466.
2. De La Roche, A. I. Plant Physiol. 1979, 63, 5-8.
3. McElhaney, R. N. J. Supramolecular Structure 1974, 2, 617-628.
4. Kasai, R.; Kitajima, Y.; Martin, C. E.; Nozawa, Y.; Skriver, L., and Thompson, Guy A., Jr. Biochem. 1979, 15, 5228-5233.
5. Kitajima, Y. and Thompson, G. A., Jr. J. Cell Biol. 1977, 72, 744-755.
6. Monero, R. D.; Blou, B.; Farias, R. N., and Trucco, R. E. Biochim. Biophys. Acta. 1973, 282, 157-165.
7. Waring, A. J. and Laties, G. G. Plant Physiol. 1977, 69, 11-16.
8. Waring, A. J.; Breidenbach, R. W., and Lyons, J. M. Biochim. Biophys. Acta. 1976, 443, 157-168.
9. Sikorska, E., and Kacperska, A. In: Plant Cold Hardiness and Freezing Stress. Mechanisms and Crop Implications, vol. 2 eds. P.H.Li and A. Sakai. Academic Press 1982, p. 261-272.
10. Hilton, J. L.; St. John, J. B.; Christiansen, M. N., and Norris, K. H. Plant Physiol. 1971, 48, 171-177.
11. St. John, J. B. Plant Physiol. 1976, 57, 38-40.
12. Kramer, P. J. Plant and Soil Water Relationships 1st Ed. McGraw-Hill, N.Y. 1949, pp. 227-234.
13. St. John, J. B. and Christiansen, M. N. Plant Physiol. 1976, 57, 257-259.
14. Willemot, C. Plant Physiol. 1977, 60, 1-4.
15. St. John, J. B.; Christiansen, M. N.; Ashworth, E. N.; and Gentner, W. A. Crop Sci. 1979, 19, 65-69.
16. Ashworth, E. N.; Christiansen, M. N.; St. John, J. B., and Patterson, G. B. Plant Physiol. 1981, 67, 711-715.
17. St. John, J. B.; Rittig, F. R.; Ashworth, E. N.; Christiansen, M. N. in "Advances in Pesticide Science" Geissbuhler, H., Ed.; Pergamon Press: New York, 1978; pp. 271-273.
18. Bartlett, G. R. J. Biol. Chem. 1959, 234, 466-468.

RECEIVED February 22, 1984

20

Bioregulators and Rubber Synthesis in the Guayule Plant

H. YOKOYAMA, W. J. HSU, E. HAYMAN, and S. POLING

Agricultural Research Service, Fruit and Vegetable Chemistry Laboratory, U.S. Department of Agriculture, Pasadena, CA 91106

Bioregulators caused a significant increase in the synthesis of rubber in the guayule plant. This was accomplished without altering the microstructure (*cis*-1,4 isoprene) of the polyisoprene molecule. The carbon-13 nuclear magnetic resonance spectra show a complete absence of signals attributable to structural isomers and confirm that guayule rubber from treated plants is a highly stereospecific polymer composed entirely of *cis*-1,4 isoprene units. The gel permeation chromatography results suggest that these bioregulators induce new rubber molecules rather than chain extension of rubber molecules at the surface of existing rubber particles.

Natural polyisoprene rubber is distributed widely in the plant kingdom. About 2000 species of plants are known to synthesize rubber. However, only a few ever produce rubber in substantial amount for commercial use. Two of these, the rubber tree *Hevea brazíliensis*, and the guayule shrub *Parthenium argentatum*, have been continuing sources of natural rubber. These two plants have contrasting climatic requirements. *Hevea* is native to equatorial lowland rainforests in the Amazon basin. On the other hand, the guayule plant grows wild in the upland plateau areas of Mexico and Texas with subtropical-temperate climates and meager rainfall. Despite these differences, the two plants produce a similar rubber. The physical, chemical, and mechanical properties of *Hevea* rubber and guayule rubber are similar (1).

Guayule Plant

The guayule plant is a member of the sunflower family *Compositae* and belongs to the genus *Parthenium*. The guayule plant is *Parthenium argentatum*, so designated because of a silvery sheen on its grey-green foliage, and is one of 16 species of *Parthenium*. It is the only *Parthenium* species known to produce rubber in any appreciable quantity (2). Unlike the rubber in *Hevea* and other latex-producing plants in the family Euphorbiaceae, rubber in the guayule plant is not contained in ducts but in single thin-walled parenchyma cells of the cortex, pith, and vascular rays of the stems and roots, and to a much lesser extent in the leaves (3). The rubber appears as membrane bound particles (4). A bushy perennial shrub, guayule, could be cultivated in its native habitat and in the warmer areas of California, Arizona, and New Mexico. It could also be cultivated in the subtropical-temperate regions of Asia, Australia, and Africa.

Plant Bioregulation

The single major obstacle to commercial production of guayule rubber is the low yield. Thus, this has stimulated research towards increasing yield from this plant. Traditional breeding programs and a variety of more exotic hybridization techniques may not be the only ways to increase the yield of rubber in the guayule plant. A promising approach to improving the yield characteristics is through bioregulation of the synthesis of polyisoprenes to cause an accumulation of increased amounts of rubber. This approach is based on the discovery of bioregulators that stimulate the production of additional quantities of tetraterpenoids in plant tissues (5).

Bioregulators

A series of bioregulators were developed for possible use in influencing rubber formation in guayule. Thus far, one of the most effective bioregulators is DCPTA (2-diethylaminoethyl-3,4-dichlorophenylether) as shown in Table I (6).

Table I. Bioinduction of Rubber in Stem and Branch Tissues of Guayule (18 months old) by 2-Diethylaminoethyl-3,4-Dichlorophenylether. The Plants were Treated with 2000 ppm of Bioregulator, 1000 ppm Isopropanol, and 500 ppm Ortho-X77. All Plants were Harvested 40 Days after Treatment. Each Result Represents the Mean of 6 Plants.

Strain	Rubber Content (mg/g dry wt.)	
	Control	Treated
212	58	146
228	61	100
230	42	77
234	52	92
239	57	122
241	48	122
242	51	108

A number of benzylalkylamines and benzylfurfurylamines were synthesized (7). p-Bromobenzylfurfurylamine and N-methylbenzylhexylamine caused marked increases in the content of rubber in guayule (Tables II and III).

Table II. Bioinduction of Rubber in Stem and Branch Tissues of Guayule (18 months old) by p-bromobenzylfurfurylamine. The Plants were Treated with 2000 ppm of Bioregulator, 1000 ppm of Isopropanol and 500 ppm of Ortho-X77. All Plants were Harvested 40 Days after Treatment. Each Result Represents the Mean of 6 Plants.

Strain	Rubber Content (mg/g dry wt.)	
	Control	Treated
228	58	146
239	52	121
234	54	162
89	72	207

Table III. Bioinduction of Rubber in Stem and Branch Tissues of Guayule (10 months old) by N-methylbenzylhexylamine. The Plants were Treated with 2000 ppm of Bioregulator, 1000 ppm of Isopropanol and 500 ppm of Ortho-X77. All Plants were Harvested 40 Days after Treatment. Each Result Represents the Mean of 5 Plants.

Strain	Rubber Content (mg/g dry wt.)	
	Control	Treated
593	39	131

At lower levels of concentration, the bioregulators appear to have an effect on the total rubber content as shown in Table IV (8).

Table IV. Effects of Several Bioregulators on Rubber Content of Guayule

Treatment	Rubber Content[a]	%[b]
Control	15.5 ± 3.9	5.3 ± 0.7
250 ppm of 2-(3,4-dichlorophenoxy)-triethylamine	23.6 ± 6.5	5.0 ± 1.4
125 ppm of 2-(3,4-dichlorophenoxy)-triethylamine	23.5 ± 6.2	6.7 ± 0.5
500 ppm of 2-(2,4-dichlorophenoxy)-triethylamine	24.3 ± 4.2	6.0 ± 1.5
1000 ppm of N-methylbenzylhexyl-amine	26.4 ± 7.4	6.6 ± 0.2
500 ppm of 2-(3,4-dimethylphenoxy-triethylamine	29.6 ± 8.4	6.6 ± 0.7

[a] Grams of rubber per plant ± the standard deviation for four to six plants. [b] Percent of rubber ± the standard deviation for four to six plants.

The 2,4-dichloro and 3,4-dimethyl analogs of DCPTA appeared to increase the total rubber content of the plants.

Mode of Action of Bioregulators

The effect of bioregulators on the increases in total yield of rubber per plant is limited to a great extent by the availability of storage areas for the newly synthesized rubber molecules. Studies on native guayule and some hybrids of guayule and other *Parthenium* species have shown greater variability of amounts of parenchyma tissues (which act as storage areas for rubber) present in the plants (9). Studies also showed that newly induced rubber molecules formed after the young plant was treated with the bioregulator *p*-bromobenzylfurfurylamine are stored in the parenchymatous cells which did not have a notable quantity of rubber before the treatment (10). In these studies, the bioinduction of rubber was observed over a period of time (55 days) by taking tissue slices of the stems at intervals of 0, 13, 24, 34, and 55 days after treatment. Micrographs taken at 24 and 34 days showed significant increases in rubber content; the dark areas stained with Sudan black in Figures 1, 2, and 3. No increases in rubber content were noted between 0 and 13 days and between 34 and 55 days. Bioinduction appeared to cease after all of the cells which are capable of

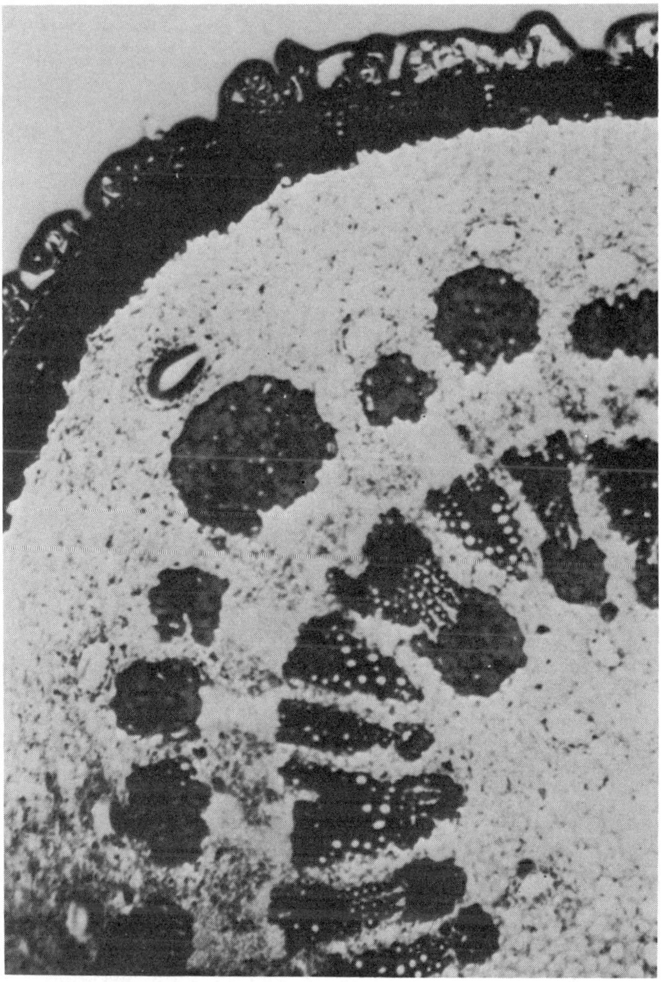

Figure 1. Cross section of a young guayule stem 13 days after treatment with p-bromobenzylfurfurylamine. No increase in rubber content (dark spots) is observed when compared to cross section at 0 days. Cross section at 0 day is identical to one at 13 days.

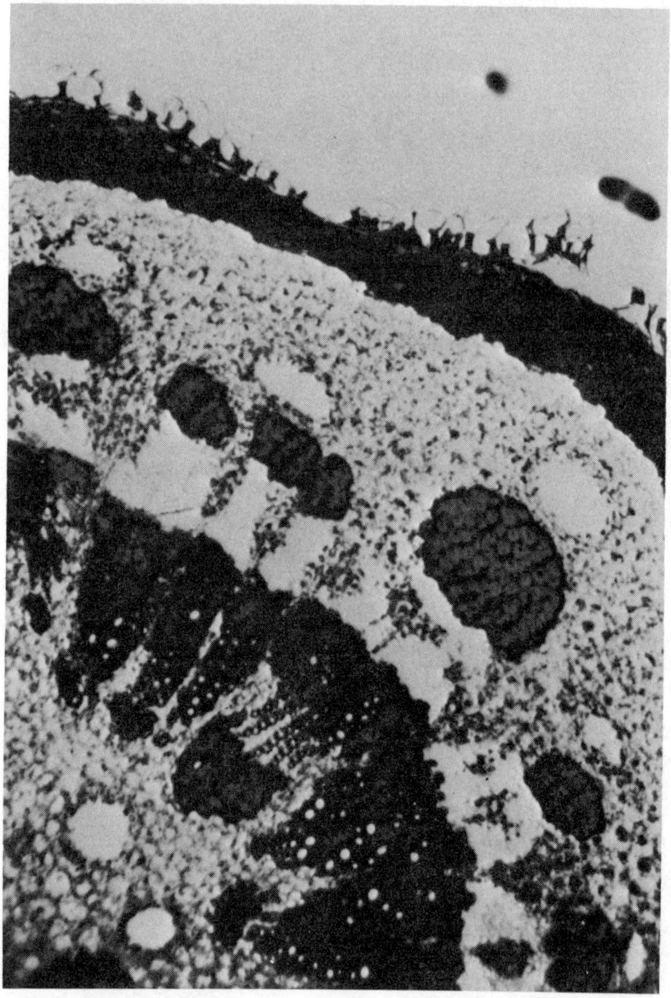

Figure 2. Rubber content (dark spots) 24 days after treatment. Cross section of young stem show rubber content increasing in response to application of bioregulator p-bromobenzylfurfurylamine.

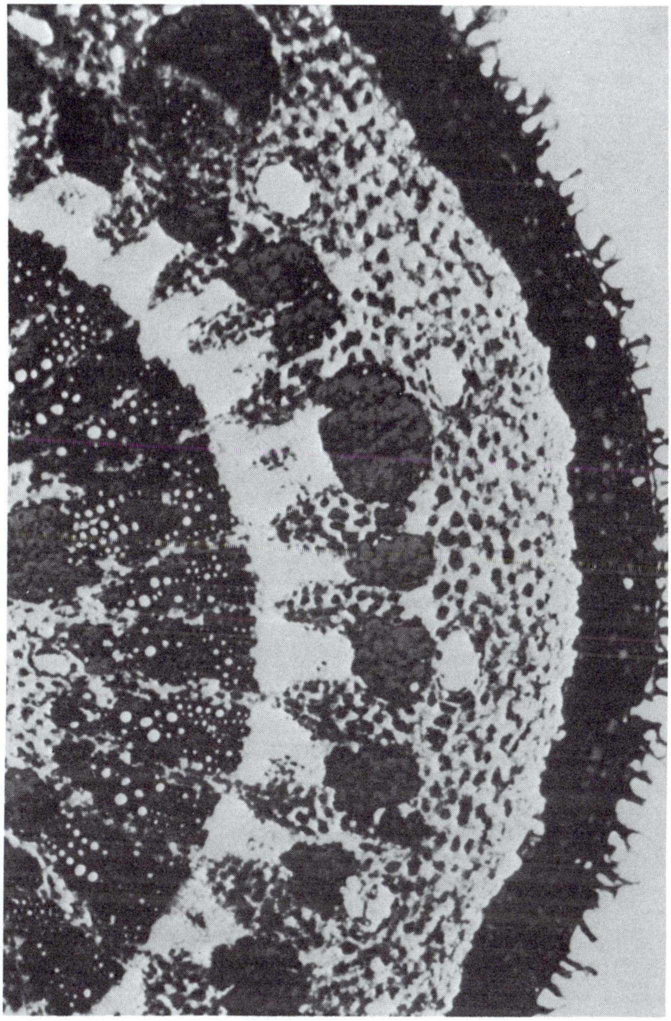

Figure 3. Rubber content (dark spots) 34 days after treatment with _p_-bromobenzylfurfurylamine. Cross section of young stem at 55 days after treatment shows no increase in rubber content.

producing and storing rubber have been filled with rubber. Approximately a three-fold increase in the number of rubber particles was seen.

Findings thus far indicate that the guayule plant has a certain capacity to form and store rubber, and this is dependent on the particular strain. Bioinduction will not be feasible beyond the biological potential unless this potential is increased by cell differentiation or enlargement. The levels of several enzymes involved in the synthesis of cis-polyisoprene and trans-farneylpyrophosphate (FPP) from mevalonic acid (MVA) in the stems of DCPTA-treated and control plants are shown in Table V (11).

Table V. Effect of DCPTA Treatments on the Activity of Enzymes Involved in the Synthesis of cis-Polyisoprene and FPP from MVA in Guayule stems.

The enzymic activities are averages of two replicates each.

Enzyme	DCPTA			
	Control	Treated	Control	Treated
	nmol mg^{-1}	Protein h^{-1}	nmol g^{-1}	Fresh Wt h^{-1}
MVA kinase	43.8	66.7	233.9	439.0
IPP isomerase	4.1	7.5	19.8	39.8
Rubber transferase	4.5	9.1	21.9	48.3
FPP synthetase	6.4	20.2	31.5	107.3

After 120 days of growth following DCPTA application, the stems of treated plants contained about a 2-fold increase in rubber (Table VI) and a 1.5-fold greater MVA kinase activity than did the control plants. The activities of isopentenylpyrophosphate (IPP) isomerase and rubber transferase were doubled in the stems of the DCPTA-treated plants and there was about a 3-fold stimulation of FPP synthetase in the bioregulator-treated guayule plants.

Rubber Quality

Like Hevea rubber, guayule rubber is a polymer of the simple 5-carbon isoprene molecule and has the cis-1,4 shape (Figure 4). The isoprene units are joined together end-to-end to form a giant molecule containing tens of thousands of carbon atoms in a linear chain identical to that of Hevea rubber and with similar molecular weight (12).

Any improvement of yield of rubber in guayule must be accomplished without diminishing the physico-chemical characteristics of rubber. The precise stereochemistry must be maintained unaltered. These are essential requirements for commercialization in the bioinduction of rubber formation. Rubber samples isolated from guayule plants treated with DCPTA were compared to those from untreated plants as well as Hevea rubber. Carbon-13 nuclear magnetic resonance (NMR) spectra (Figure 5) are identical and confirm the structural and geometrical purity of guayule rubber isolated from treated plants. The NMR spectra show a complete absence of signals attributable to stereo or structural isomers, namely trans-1,4 isoprene units, demonstrating that guayule rubber from treated plants is a highly stereospecific polymer composed entirely of cis-1,4 isoprene units. The NMR spectra confirm that improvement of rubber yield is accomplished without altering the microstructure of the rubber. The molecular weight distribution of guayule rubber and Hevea rubber was examined by gel permeation chromatography (GPC) as shown in Figure 6. The distribution of molecular weights of the polyisoprene chain in rubber from untreated plants is identical to that from treated plants and both are unimodal (Figure 7). The GPC results suggest that these bioregulators induce new rubber molecules rather than a chain extension of rubber molecules at the surface of existing rubber particles.

Table VI. Effect of DCPTA on the Accumulation of Rubber in Guayule Plants.

The data are the average of three replicates. Data from 120 days are averages of two replicates each.

Treatment	Induction Period	Dry Wt	Rubber Content	Percent Rubber
	d	g	mg/plant	%
Control	30	5.21	15.9	0.30
DCPTA	30	6.61	20.5	0.31
Control	60	11.0	76.0	0.69
DCPTA	60	11.5	125.0	1.04
Control	90	13.4	168.0	1.25
DCPTA	90	14.4	379.0	2.63
Control	120	13.3	377.0	2.83
DCPTA	120	14.3	748.0	5.23

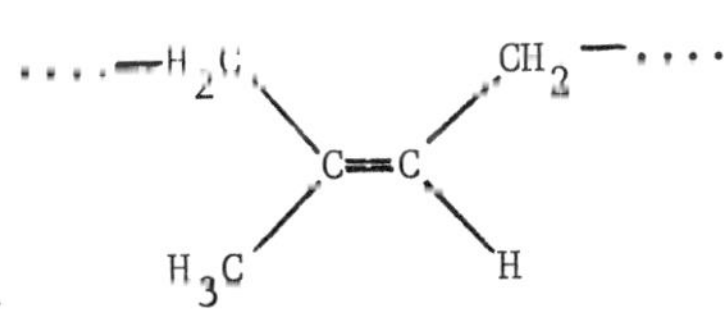

Figure 4. Simple 5-carbon isoprene unit of rubber.

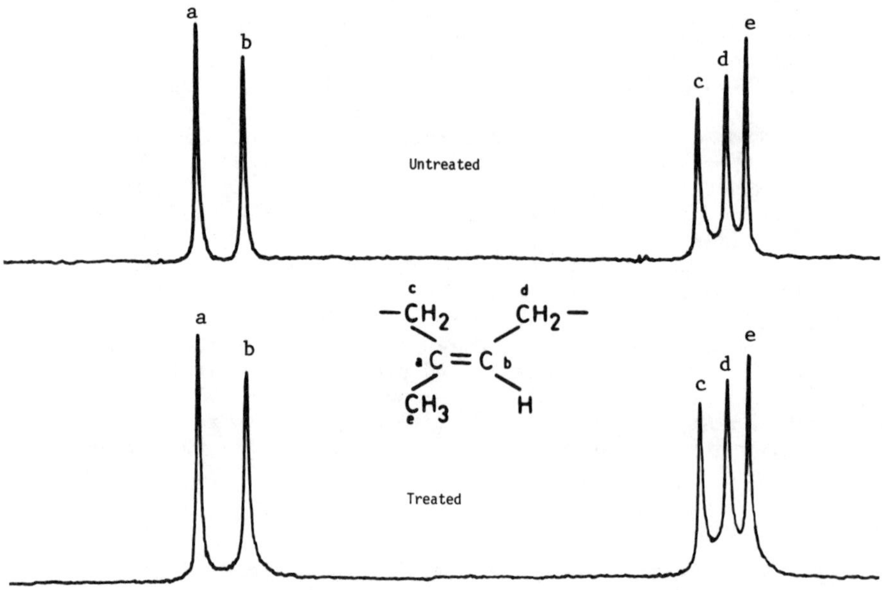

Figure 5. Comparison of C-13 nuclear magnetic resonance spectra of guayule rubber from untreated plants and plants treated with DCPTA (2-diethylaminoethyl-3,4-dichlorophenylether).

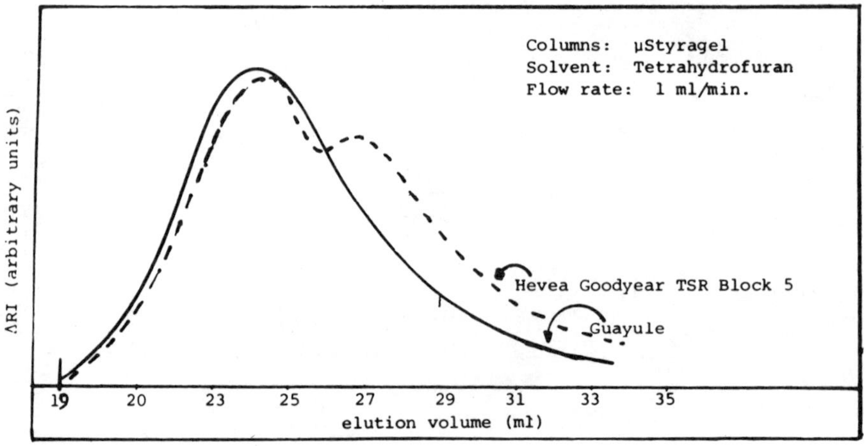

Figure 6. Gel permeation chromatography (GPC) of guayule rubber and Hevea rubber. The GPC analysis was conducted using a Waters model 6000 chromatography with differential refractive index detector, solvent system of tetrahydrofuran (at 28°C) stabilized with 200 ppm of 2,6-di-tert-butyl-4-ethylphenol and a set of 5 μ-Styragel columns each with a normal porosity of 10^7, 10^6, 10^5, 10^3, and 500Å.

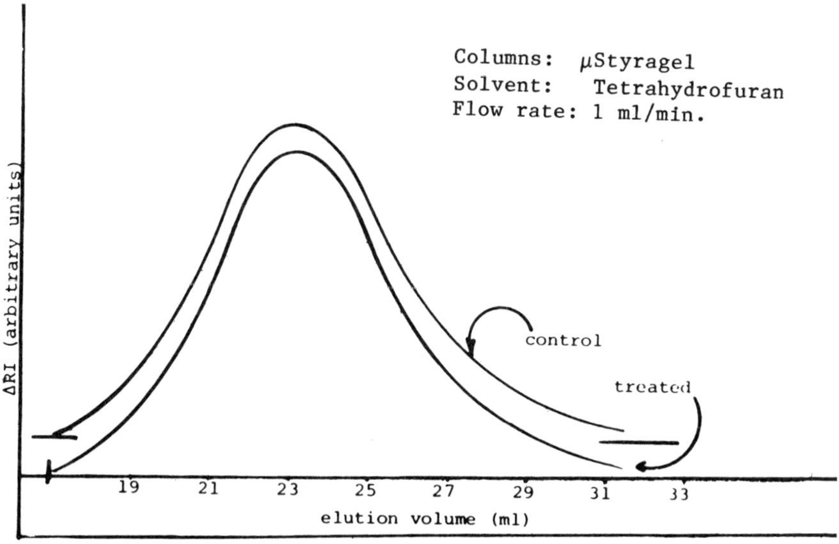

Figure 7. GPC comparison of guayule rubber from untreated plants and plants treated with DCPTA.

Reference to a product name or company does not imply endorsement of that product or company by the U.S. Department of Agriculture to the exclusion of others that may be available.

Literature Cited

1. McIntyre, D. In "Guayule"; Guzman, W., Ed., Consejo Nacional de Ciencia y Technologia: Mexico, 1978; Chap. 2
2. Hammond, B. L., Polhamus, L. G. U.S. Department of Agriculture Technical Bulletin No. 1329, 1965; pp. 5-7.
3. Curtis, O. P. Plant Physiology 1947, 22, 33.
4. Gilliland, M. G.; Van Staden, J. Z. Pflanzenphysiol. Bd. 1983 110, 285.
5. Yokoyama, H., Hsu, W. J., Poling, S., Hayman, E. Proc. Int. Soc. Citriculture 1977, p. 717.
6. Yokoyama, H.; Hayman, E., Hsu, W. J., Poling, S.; Bauman, A. Science 1977, 197, 1076.
7. Poling, S.; Hsu, W. J., Yokoyama, H. Phytochemistry 1980, 19, 1677.
8. Hayman, E., Yokoyama, H., Gold, S. J. Agric. Food Chem. 1983, 31, 1120.
9. Mehta, I. Amer. J. Botany 1982, 69, 502.
10. Bauer, T., Glaeser, R.; Yokoyama, H. Unpublished data.
11. Benedict, C. R., Reibach, P. H., Madhavan, S., Stipanovic, R. V., Keihly, J. H., Yokoyama, H. Plant Physiol. 1983, 72, 897.
12. Campos-Lopez, E., Palacios, J. J. Polymer Sci. 1976, 14, 15.

RECEIVED April 20, 1984

21

Induction of Phytoalexin Synthesis in Plants Following Microbial Infection or Treatment with Elicitors

JÜRGEN EBEL

Biologisches Institut II, Universitat Freiburg, D-7800 Freiburg, Federal Republic of Germany

Plants are exposed to a large number of microorganisms in their natural environment. Although severe damages may be caused by microbial pests under unfavorable conditions, normally plants successfully resist attempted infections. Effective defence mechanisms have evolved in plants which secure their survival in the presence of these challenges. It is now evident that resistance mechanisms of plants are effective at different levels in host-parasite interactions including preformed physical and chemical defence barriers as well as defences triggered by the invader (1). These mechanisms probably often act coordinately during restriction of microbial invasion.

Phytoalexin Theory of Disease Resistance

One type of active response of plants to invading microorganisms is the production of low molecular weight antimicrobial compounds called phytoalexins. Originally, the "phytoalexin concept" was developed by Müller and Börger (2) from their studies on fungal infections. Since then it has been well documented that phytoalexins are produced by plants not only in response to interactions with fungi, but also in response to bacteria, viruses, and nematodes (3, 4). In view of these findings phytoalexins are now defined as "low molecular weight antimicrobial compounds that are both synthesized by and accumulate in plants after their exposure to microorganisms" (4).

Phytoalexins constitute a chemically heterogeneous group of substances which includes various classes of natural products. There are about 100 known phytoalexins distributed among isoflavonoids, sesquiterpenes, diterpenes, polyacetylenes, dihydrophenanthrenes, and stilbenes. The structures of a selected number of characteristic examples are shown in Figure 1. It seems likely that other classes of natural products will be implicated to be phytoalexins as the scope of investigations is broadened. For example, recently, a styrylbenzoxazinone, avenalumin I, and two structurally related

0097-6156/84/0257-0257$06.00/0

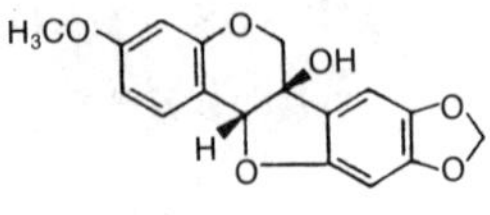

1

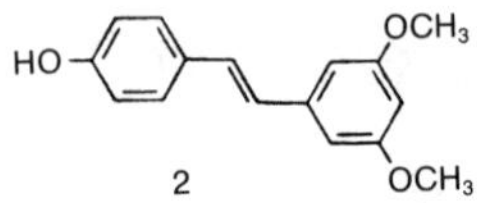

2

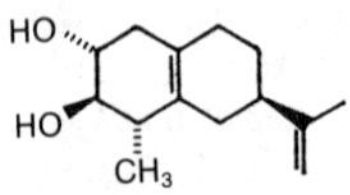

3

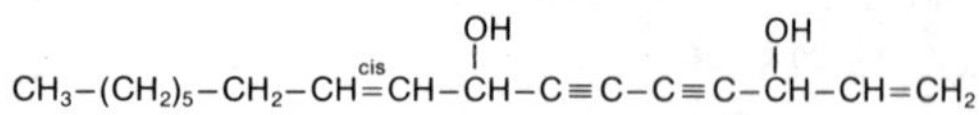

4

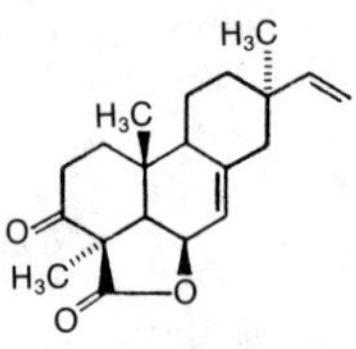

5

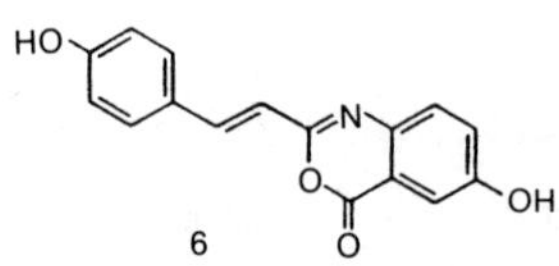

6

Figure 1. Structures of a number of selected phytoalexins. Key: 1, (+) pisatin (pterocarpan) from pea (Pisum sativum); 2, pterostilbene (stilbene) from grape (Vitis vinifera); 3, rishitin (sesquiterpene) from potato (Solanum tuberosum); 4, falcarindiol (polyacetylene) from tomato (Lycopersicum esculentum); 5, momilacton A (diterpene) from rice (Oryzasativa); 6, avenalumin I (styrylbenzoxazinone) from oat (Avena sativa).

compounds, have been identified from oat (Avena sativa) following infection with Puccinia coronata f. sp. avenae (5).

A comprehensive survey of the nature and distribution of phytoalexins in the plants kingdom does not yet exist but generally the compounds produced by different members of a plant family are structurally related (3, 4). The legumes produce isoflavonoids, the solanaceae sesquiterpenes, the compositae polyacetylenes, the orchidaceae dihydrophenanthrenes, and the vitaceae stilbenes. Chemical diversity within a plant family, and also within a species, such as the formation of the furanoacetylene, wyerone acid, and the isoflavonoid, medicarpin, in broad bean (Vicia faba) (6), or the polyacetylenes, falcarinol and falcarindiol, and the sesquiterpene, rishitin, in tomato (Lycopersicum esculentum) (7, 8) is less frequently found. Thus, it has been proposed to use phytoalexin induction as a new approach for chemotaxonomic studies with higher plants (9).

Extensive research on certain species in the leguminosae and solanaceae has revealed that a plant may produce several structurally related phytoalexins as is the case with green beans (Phaseolus vulgaris), soybeans (Glycine max), and potatoes (Solanum tuberosum). The major isoflavonoid phytoalexins of soybean which comprise several biosynthetically interrelated pterocarpans (10-12) are shown in Figure 2.

The biological activity of the phytoalexins has been extensively studied in vitro using different types of bioassays. In general, phytoalexins are toxic to fungi, bacteria, higher plant cells, and also animal cells. Biocidal as well as biostatic properties have been observed. Effective doses for growth inhibition of fungi and bacteria fall within one order of magnitude (10^{-5} to 10^{-4} mol/l). Individual organisms may express differential sensitivity towards a given phytoalexin. In some cases the ability to metabolize and thereby detoxify phytoalexins might be related to the pathogenicity of a particular organism (13). The mechanisms by which phytoalexins express their toxic effects are not well understood. Considering the great diversity of phytoalexin structures, a single mode of action is unlikely. It has been proposed that phytoalexins are multi-site toxicants causing dysfunction of various membrane-linked processes by a general disruption of membrane integrity (14).

This view is supported by recent studies on the mode of action of the soybean phytoalexin, glycinol (15, 16). However, mechanisms of action at specific sites have also been suggested (17, 18).

Many observations demonstrated that the same antimicrobial compounds as those formed during host-parasite interactions are also produced following treatment of plants with various chemicals, irradiation by ultra-violet light, exposure to cold, and to products of microbial metabolism (3). Compounds of microbial origin which induce phytoalexin accumulation have been called elicitors (19). In addition, elicitors have been divided into those of a biological origin, biotic elicitors, and those of chemical and physical nature, abiotic elicitors (20). It is not known whether this division is of any biological significance. In those cases where compounds identical to known phytoalexins are formed in response to abiotic agents or physical treatments the compounds have repeatedly been referred to

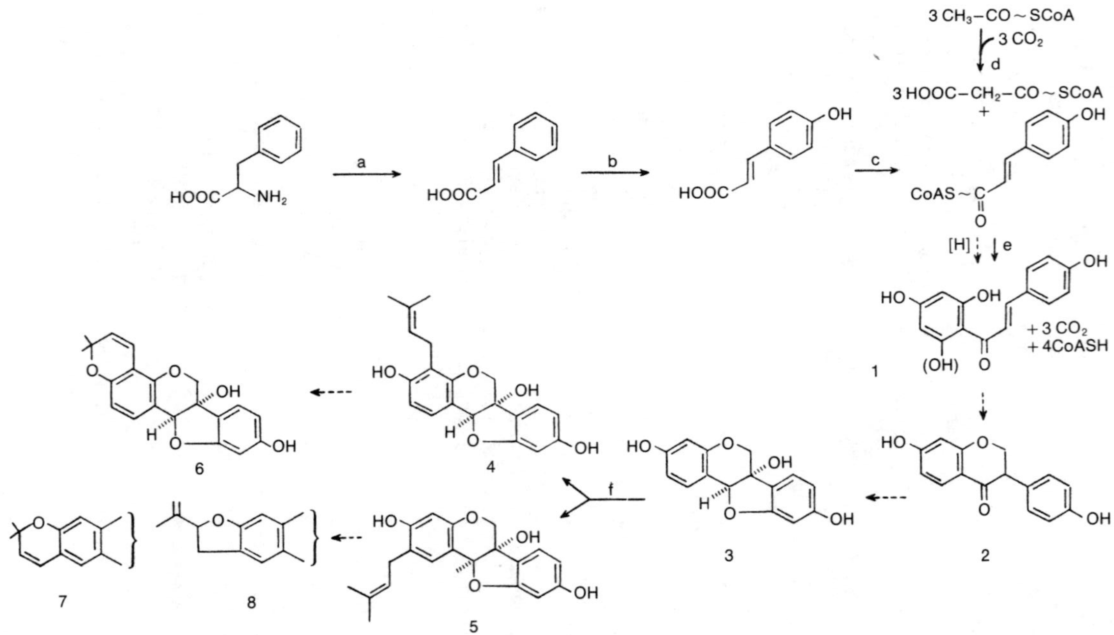

Figure 2. Proposed biosynthetic pathway of the soybean phytoalexin, glyceollin. Key: 1, 4,4',6'-trihydroxychalcone; 2, dihydroisoflavanone; 3, glycinol (3,6a,9-trihydroxypterocarpan); 4,5, glyceollidin I and II (4- and 2-dimethylallyl-trihydroxypterocarpan); 6, glyceollin I; 7, glyceollin II; 8, glyceollin III. Key to enzymes: a, phenylalanine ammonia-lyase; b, cinnamate 4-hydroxylase; c, 4-coumarate: CoA ligase; d, acetyl-CoA carboxylase; e, chalcone synthase; f, prenyltransferase.

as stress metabolites (21). In this broader context, phytoalexins may be considered as a special class of stress metabolites which, however, have acquired special importance because of their proposed role in disease resistance.

Many recent reports have been concerned with the role of phytoalexins in determining disease resistance of plants. The evidence in support of such a role is circumstantial. The majority of data comes from plant-fungus interactions in which resistance is expressed following penetration and is associated with the necrosis of plant cells (hypersensitive reaction). The evidence for a role in resistance to bacterial or viral diseases is less clear. There is a need for future critical investigations of the natural sequence of events during expression of resistance to further examine the hypothesis that phytoalexins are important factors in defence. This includes the questions of the antimicrobial activity of phytoalexins *in vivo*, their rates of accumulation, and their localization at the micro-sites around the invading microorganisms.

The rapid expansion of the literature on phytoalexins reflects a steadily growing interest in research of plant-parasite interactions. A number of recent reviews on various aspects of the subject are available (3, 4, 22-24). In this contribution, studies on the induced synthesis of phytoalexins in soybean following fungal infection or treatment with a fungal elicitor will be discussed in more detail.

Induced Phytoalexin Synthesis in Soybean

Phytophthora megasperma f. sp. *glycinea* (25), a fungus causing stem and root rot of soybean (*Glycine max*), exists as several physiological races which can be differentiated by their reaction with various cultivars of soybean (26). In general, the combination or interaction between the host plant and the pathogen is termed as "compatible" if the host is susceptible and the pathogen is virulent. The combination in which the host is resistant and the pathogen is avirulant is accordingly described as "incompatible". Susceptible soybean hypocotyls inoculated with mycelium or zoospores of *P. megasperma* develop water-soaked, spreading lesions within 24 h after inoculation. Resistance is expressed in certain race-cultivar combinations by the equally rapid appearance of dark brown lesions which remain localized to inoculation sites. In such an interaction, invasion by the fungal hyphae causes rapid death of epidermal and neighbouring cells and fungal growth is restricted to the necrotic lesions. The resistance response has been described as hypersensitivity. Following infection with an incompatible race of the fungus, the levels of the soybean phytoalexins (Figure 2) can increase from below detectable limits to greater than 10% of the dry weight of the infected soybean tissue within 24 to 48 h after inoculation (27). This implies the existance of an effective set of regulatory controls which operate during the host-parasite interaction leading to phytoalexin accumulation. A number of important questions arise as to the nature and action of the regulatory controls at the molecular level. Answers to these questions will undoubtedly advance our understanding of the role of phytoalexins in disease resistance as well as of the control of gene expression in plant cells.

The process of induction of phytoalexin synthesis in plant tissues can be viewed as a signal-response event in which the challenging organism generates an external signal that finally is transformed into a response of the plant cell. Many investigations were aimed at identifying the chemical nature of the signal(s) produced by microorganisms which might be responsible for inducing phytoalexin synthesis (28, 29). Several types of fungal and bacterial compounds including polysaccharides, polypeptides, glycoproteins and lipid substances have been shown to serve as elicitors of phytoalexin synthesis when applied to various plant tissues. The best characterized elicitor of *P. megasperma* is a structural polysaccharide found in the cell walls of the fungus. The nature and activity of this polysaccharide elicitor was extensively studied by Ayers et al. (28, 30-33) employing various methods of fractionation and a number of biological assays. Several properties of the biologically active material including size heterogeneity, stability to heat and pH extremes, insensitivity to digestion by proteolytic enzymes, and structural analysis revealed that the polysaccharide elicitor is a 1,3-glucan with glucosyl branches at position 6 of some of the residues. The presence of the ß-anomeric configuration was indicated by the sensitivity of the branched 1,3-glucan to digestion with an exo-ß-1,3-glucanase, and was confirmed by optical rotation and NMR studies (28, 32). The conclusions derived from the earlier studies were strongly confirmed by a recent thorough structural analysis of a small, biologically active heptaglucosyl fragment and by chemical synthesis of the active heptasaccharide (34,35).

Some aspects of the biological specificity of the *P. megasperma* glucan elicitor have been evaluated. Elicitor preparations from both compatible and incompatible races of the fungus were quantitatively identical in their abilities to induce glyceollin accumulation in the same soybean cultivar (32). Therefore, it was concluded that the elicitors by themselves are insufficient to account for the observed specificity of the interaction between the different races of *P. megasperma* and various soybean cultivars. The glucan elicitor from either race of the fungus, however, was capable of protecting soybean hypocotyls from symptom expression by a normally compatible race of *P. megasperma* if the elicitor was applied to the hypocotyls 6 h before challenge with the fungus (28). Further investigations on the important question of the role of elicitors in determining the specificity of compatible and incompatible interactions in different race-cultivar combinations were recently reported by Ziegler and Pontzen (36). These authors demonstrated that the glucan elicitor induced accumulation of glyceollin in soybean cotyledons was suppressed by a mannan-glycoprotein from a compatible, but not from an incompatible race of *P. megasperma*. It is not yet known, whether interference of the mannan-glycoprotein from a compatible race with a normally incompatible host-fungus interaction would convert this into a compatible interaction. Other components of the *P. megasperma* culture fluid have been described to protect soybean seedlings from infection with compatible races to variable extents (37, 38). The role of suppressors and these other factors as determinants of specificity needs further exploration.

It seems likely that plant enzymes released during plant-fungus interaction are involved in the partial degradation of fungal cell walls and the release of soluble cell wall elicitors. A number of ß-glucan hydrolases from plants are known which could serve this function (39-44). Keen et al. (43, 44) reported that an endo-ß-1,3-glucanase from soybean cotyledons liberated elicitor-active fragments following treatment of P. megasperma hyphal cell walls with the purified enzyme. Other investigations indicated that microbial (29, 45) or even plant (46) enzymes might be active in the release of constitutive (47) or endogenous (48) elicitors of plant origin. Recent reports which showed that the endogenous soybean elicitor is a pectin fragment of the plant cell walls (48, 49) and that a pectin fragment of the plant cell walls might be involved in elicitation of casbene in castor bean (Ricinus communis) by a fungal endopolygalacturonase (50) are in agreement with this hypothesis.

Soybean tissues including hypocotyls, cotyledons, and cultured cells accumulate a number of structurally related phytoalexins (10-12) as depicted in Figure 2 following inoculation with P. megasperma or treatment with a glucan elicitor derived from the cell walls of the fungus. The predominant compound in all tissues is glyceollin I with lower amounts of the structural isomers glyceollin II and III co-occuring (51, 52). In this review, glyceollin will be a designation collectively referring to the mixture of the three structural isomers. Glycinol, a biosynthetic precursor of glyceollin, is a major component of the induced substances isolated from cotyledons, but only a minor one from those of hypocotyls and cultured cells (15, 52, 53). Several authors have investigated the relative rates of synthesis and degradation of the soybean phytoalexins to determine the mechanism of net accumulation of these compounds in response to different stimuli. Yoshikawa et al. (20, 54) suggested from their results that the enhanced accumulation of glyceollin following infection of hypocotyls with P. megasperma or treatment of cotyledons with $HgCl_2$ was due to a decrease in the rate of degradation, whereas the induced accumulation in response to cell walls of P. megasperma was caused by an enhanced rate of synthesis. Moesta and Grisebach (51, 53), however, using refined techniques for pulse-chase experiments and phytoalexin isolation obtained different results. They conclusively demonstrated that accumulation of glyceollin in response to both infection and different types of elicitor was predominantly regulated by changes in the rate of synthesis.

Induction of Enzymes of Phytoalexin Synthesis in Soybean

The pterocarpanoid phytoalexin, glyceollin, belongs to the large group of isoflavonoids which are common natural constituents of the legumes. Our present knowledge of isoflavonoid biosynthesis is based on a combination of results from radioactive tracer studies in vivo (55) and data obtained at the enzyme level in vitro (56). As outlined in Figure 2, glyceollin biosynthesis includes a number of reactions catalyzed by enzymes of general phenylpropanoid metabolism (phenylalanine ammonia-lyase, cinnamate 4-hydroxylase and 4-coumarate:CoA ligase) and of flavonoid biosynthesis (acetyl-CoA carboxylase and chalcone synthase). The chalcone synthase identified so far from different soybean tissues catalyzes the formation of 4,2',4',6'-tetrahydroxychalcone. It is not yet known how the synthesis of 4,4',6'-trihydroxychalcone, the putative precursor of glyceollin, is accomplished *in vivo* (52). Re-arrangement of the chalcone yields the isoflavonoid skeleton and further transformation the intermediate, glycinol. Recent investigations demonstrated a dimethylallylpyrophosphate:3,6a,9-trihydroxypterocarpan (glycinol) dimethylallyltransferase in a particulate fraction from elicitor-treated or infected tissues of soybean (57, 58). The major product of dimethylallyltransfer *in vitro* was glyceollidin II. The minor product was the 4-substituted pterocarpan, glyceollidin I. Both products are likely intermediates in the biosynthesis of the various structurally related soybean phytoalexins.

The accumulation of glyceollin in fungus-infected or elicitor-treated soybean tissues coincides with large transient increases in the activities of all of the above mentioned enzymes related to the biosynthesis of the pterocarpanoid phytoalexin (33, 58-62). More recently, further details of enzyme induction following treatment of cultured soybean cells with the glucan elicitor were established (52, 62). For two of the induced enzymes, phenylalanine ammonia-lyase and chalcone synthase, the changes in catalytic activities were correlated with corresponding large changes in their rates of synthesis, as determined *in vivo* and *in vitro*. The studies using a mRNA-dependent *in vitro* translation system revealed that the time courses and the degrees of elicitor-induced changes in mRNA activities for both enzymes were very similar with respect to each other (Figure 3). Following the onset of induction, the two mRNA activities increased significantly at 3 h, reached highest levels at 5 to 7 h, and subsequently returned to low values at 10 h. The changes in mRNA activities *in vitro* were closely correlated with the changes in the rates of enzyme synthesis measured *in vivo* in elicitor-treated soybean cells (62) or fungus-infected soybean seedlings (61) and preceded transient enhancements in the catalytic activities of the two enzymes (Figure 3). The results are in agreement with the assumption that the elicitor-induced increases in the catalytic activities of phenylalanine ammonia-lyase and chalcone synthase are caused by changes in the mRNA activities encoding these enzymes. Largely similar conclusions were drawn for elicitor induction of enzymes related to phenylpropanoid biosynthesis in other systems (63, 64).

Similar degrees of induction of mRNA activities and the catalytic activities of phenylalanine ammonia-lyase and chalcone synthase were observed in soybean cells in response to three diverse microbial compounds, the glucan elicitor from *P. megasperma*, xanthan, an extracellular polysaccharide from *Xanthomonas campestris*, and endopolygalacturonase from *Aspergillus niger*. However, whereas the glucan elicitor induced the accumulation of large amounts of glyceollin in soybean cells, endopolygalacturonase induced only low, albeit significant amounts, but xanthan did not enhance glyceollin accumulation under the conditions used (52). This result might imply that enzymes other than phenylalanine ammonia-lyase and chalcone synthase exert an important regulatory function in phytoalexin synthesis in soybean cells (52).

The mechanism by which the glucan elicitor causes the rapid and drastic alterations in the metabolism of soybean cells is largely unknown. An attractive hypothesis is that the primary interaction of the elicitor with the plant cell involves elicitor binding (65) at

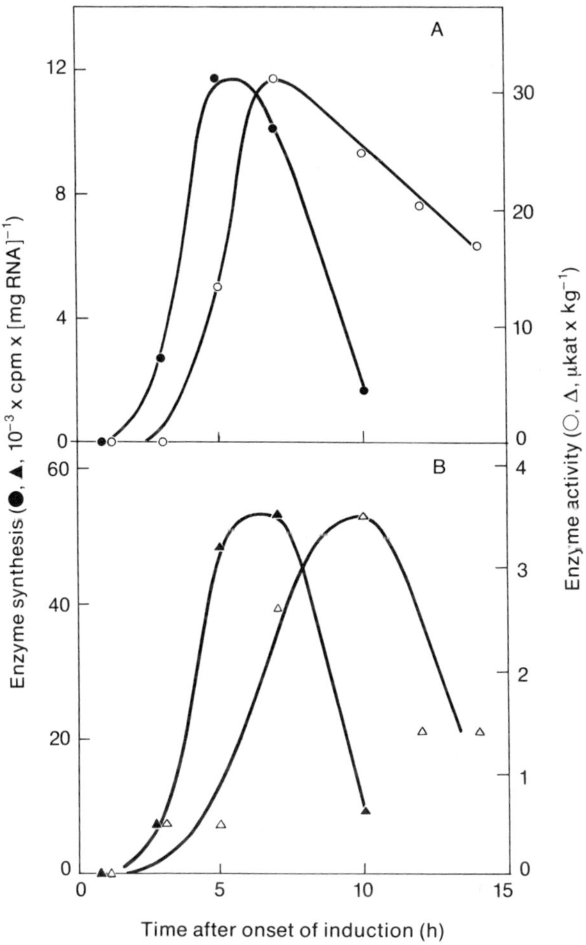

Figure 3. Changes in the rates of synthesis in vitro (●,▲) and in the catalytic activities (o,△) of phenylalanine ammonia-lyase (A) and chalcone synthase (B) upon treatment of soybean cells with glucan elicitor. Levels from non-induced control cells were substracted in all cases. Modified after Ebel et al. (52).

the cell surface. The initial interaction could generate a secondary signal which ultimately activates the expression of a number of genes related to phytoalexin synthesis. It appears unlikely that adenosine-3':5'-cyclic monophosphate which serves as a second messenger in animal systems plays an analogous role in the response of soybeans to infection with P. megasperma or the response of cultured cells to elicitor (66). However, Ca^{2+} ions could have some role in the transmissions of the primary elicitor signal into the plant cell because the presence of Ca^{2+} ions in the culture medium of soybean cells enhances the elicitor-mediated induction of phenylalanine ammonia-lyase and chalcone synthese (N. Saleh, M. Stäb, J. Ebel, unpublished). Transport studies in vitro demonstrated that the glucan elicitor inhibits active Ca^{2+} uptake by membrane vesicles of soybean cells (D. Marmé, J. Ebel, unpublished). Further investigations are required to establish whether the glucan elicitor in vivo interferes with Ca^{2+} fluxes across membranes of soybean cells and thus affects intracellular Ca^{2+} levels and consequently Ca^{2+}-mediated cellular processes which might cause the induction of phytoalexin synthesis.

Role of Phytoalexins in Determining Resistance of Soybean to P. megasperma

There is as yet no general agreement about the contribution of phytoalexins to race-specific resistance of plants (3, 4). A number of studies were aimed at determining the timing of accumulation of glyceollin and related compounds, their distribution and local concentration within soybean tissues after infection with either virulent or avirulent races of P. megasperma (51, 67-69). A few studies have been reported to correlate the changes in the amounts of glyceollin with the growth of the fungus in resistant and susceptible soybean cultivars (68-71). Some authors reported that growth differences of P. megasperma in susceptible or resistant plant cultivars were visible as early as 4 to 8 h (72) or 9 h (68) after inoculation. Beagle-Ristaino and Rissler (73) indicated that resistance in soybean roots was effective during the colonization stage of P. megasperma at 48 h when significant differences in ramification hyphal lengths in resistant versus susceptible cultivars were apparent. However, differences between the compatible and incompatible interactions are not great and may, to some extent, be influenced by seedling age (70, 71, 74), temperature (75),or inoculation technique (76). Yoshikawa et al. (68) attempted to determine the phytoalexin concentrations close to the invading fungal hyphae. They measured glyceollin levels in 0.25 mm thick freeze microtome sections and reported that concentrations at the ED_{90} value (200 μg/ml) were present at 8 h after inoculation in tissue layers of resistant soybean hypocotyls which contained advancing hyphae. The time when fungal growth ceased coincided with the time when high levels of localized glyceollin accumulation occurred. Börner et al. (69) determined the distribution of glyceollin within infected soybean hypocotyls following infection with P. megasperma. The concentration of glyceollin in the incompatible interaction reached a maximum at the infection site falling sharply towards the uninfected tissue. In contrast, glyceollin concentration in the infected compatible tissue was about one-quarter of that in the incompatible interaction and the concentration gradient towards the uninfected tissue much more shallow. In both interactions, the maximum local concentration of glyceollin 25 h after inoculation was found to be above that of the ED_{90} value. These detailed studies provide some evidence that the accumulation of glyceollin could contribute to the cessation of fungal growth in soybean hypocotyls. However, for a more critical evaluation of the significance of the soybean phytoalexins in diseases resistance, it is desirable to obtain quantitative measurements of the exact spacial and temporal distribution of glyceollin at the cellular level within soybean tissues at and near infection structures of the invading microorganism. The recent development of a radioimmunoassay for glyceollin I (77) and an immunofluorescent technique for the detection of P. megasperma hyphae in situ (78) as well as the application of the laser-microbe-mass-analysis (LAMMA) to the detection of glyceollin (79) may allow a quantitative analysis of the infection process in very small amounts of tissue.

But even with the refinement of the analytical methods, it may prove difficult to unequivocally assess the role of glyceollin as other factors may contribute to the disease resistance of soybean. In realizing this problem and being aware that there are as yet no soybean mutants available which are blocked at a specific stage in the biosynthetic pathway of glyceollin Moesta and Grisebach (80) studied the effect of in situ inhibition of glyceollin synthesis on disease development in soybean. The application of L-2-amino-oxy-3-phenylpropionic acid, a potent inhibitor of phenylalanine ammonia-lyase (81), at a concentration of about 0.5 mmol/l to soybean hypocotyls 4 h prior to infection inhibited glyceollin accumulation by more than 90% and converted the incompatible interaction with race I of P. megasperma into a compatible one. Although these results add support for a role of glyceollin in the defence of soybean against P. megasperma, it is possible that besides glyceollin accumulation the production of other fungitoxic phenylpropanoid compounds is inhibited simultaneously by the method used.

The potential use of phytoalexins as disease control agents has been suggested on several occasions (82). One potential approach involves the application to susceptible plants of chemicals that are not phytoalexin elicitors, but interfere with the plant-parasite interaction to cause higher production of phytoalexins and thus increase the resistance of the plant. The enhanced accumulation of phytoalexins in plants treated with systemic fungicides has been described by several authors (69, 83, 84).

Studies with the systemic fungicide,2,2-dichloro-3,3-dimethyl-cyclopropane carboxylic acid (WL 28 325) showed that the ability of rice (Oryza sativa) leaf tissue to produce the diterpene phytoalexins, momilactones A and B, in response to infection with the blast fungus, Pyricularia oryzae, was markedly enhanced by prior treatment of plants with the fungicide (83). The cyclopropane derivative possessed little activity against the fungus in vitro and did not act as a phytoalexin elicitor. However, the fungicide treated plants were resistant to P. oryzae. Enhanced phytoalexin accumulation has also been reported in soybean hypocotyls of a susceptible cultivar pre-treated with the systemic fungicide Ridomil (±)-1-methyl 2-N-(methoxyacetyl)-2,6-xylidino propionate] upon inoculation with P. megasperma (84)). The activity of the fungicide was also associated with a hypersensitive type of response to invading hyphae which was similar to that of an incompatible plant-parasite interaction. Börner et al. (69) demonstrated that pre-incubation of a susceptible cultivar with Ridomil (20 ppm) caused an increase in glyceollin concentration per dry weight of infected tissue but not of total glyceollin content per hypocotyl, the distribution being very similar to that found in inoculated resistant seedlings. Light and electron microscopic studies have recently shown that Ridomil concentrations as low as 0.5 ppm lead to changes in the ultrastructure of P. megasperma (85, 86). Such fungicide concentrations may occur in fungicide-treated soybean seedlings at the infection site (87) and the enhanced glyceollin concentration may therefore be caused by the release of phytoalexin elicitors from hyphae damaged by the fungicide. It is conceivable that in such a situation the restriction of disease development may be the result of both the toxicity of the applied fungicide and the induced phytoalexins.

It is apparent from the studies discussed above that the expression of resistance in plants against microbial attack involves a complex interaction of multiple mechanisms. Microbial elicitors of phytoalexin synthesis and chemicals such as fungicides which alter the balance between host plant and parasite will be valuable tools for future studies into a deeper understanding of the post-infectional events in plants which may be related to disease resistance.

Acknowledgments

The authors work is supported by Deutsche Forschungsgemeinschaft and by BASF (Ludwigshafen). I thank Drs. H. Grisebach and M. Hahn for valuable suggestions during the preparation of this manuscript.

Literature Cited

1. Horsfall, J.C.; Cowling, E.B. "Plant Disease"; Academic: New York, 1980; Vol. V.
2. Müller, K.O.; Börger, H. Arb. Biol. Reichsanst. Land-Forstwirtsch. 1939, 23, 189-231.
3. Grisebach, H.; Ebel, J. Angew. Chem. Int. Ed. Engl. 1978, 17, 635-647.
4. Bailey, J.A.; Mansfield, J.W. "Phytoalexins"; Blackie: Glasgow, London, 1982; pp. 1-334.
5. Mayama, S.; Matsuura, Y.; Iida, H.; Tani, T. Physiol. Plant Pathol. 1982, 20, 189-199.
6. Hargraeves, J.A.; Mansfield, J.W.; Rossall, S. Physiol. Plant Pathol. 1977, 11, 227-242.
7. Sato, N.; Tomiyama, K.; Katsui, N. Ann. Phytopathol. Soc. Japan 1968, 34, 344-345.
8. de Wit, P.J.G.; Kodde, E. Physiol. Plant Pathol. 1981, 18, 143-148.
9. Ingham, J.L.; Harborne, J.B. Nature 1976, 260, 241-243.

10. Burden, R.S.; Bailey, J.A. Phytochemistry 1975, 14, 1389-1390.
11. Lyne, R.L.; Mulheirn, L.J.; Leworthy, D.P. J. Chem. Soc. Chem. Commun. 1976, 497-498.
12. Lyne, R.L.; Mulheirn, L.J. Tetrahedron Letters 1978, 3127-3128.
13. Van Etten, H.D.; Matthews, D.E.; Smith, D.A. in "Phytoalexins"; Bailey, J.A.; Mansfield, J.W.,Eds.; Blackie: Glasgow, London, 1982; pp. 181-217.
14. Smith, D.A. in "Phytoalexins"; Bailey, J.A.; Mansfield, J.W., Eds.; Blackie: Glasgow, London, 1982; pp. 218-252.
15. Weinstein, L.I.; Hahn, M.G.; Albersheim, P. Plant Physiol. 1981, 68, 358-363.
16. Weinstein, L.I.; Albersheim, P. Plant Physiol. 1983, 72, 557-563.
17. Kaplan, D.T.; Keen, N.T.; Thomason, I.J. Physiol. Plant Pathol. 1980, 16, 319-325.
18. Boydston, R.; Paxton, J.D.; Koeppe, D.E. Plant Physiol. 1983, 72, 151-155.
19. Keen, N.T.; Partridge, J.E.; Zaki, A.I. Phytopathology 1972, 62, 768.
20. Yoshikawa, M. Nature 1978, 275, 546-547.
21. Stoessl, A.; Stothers, J.B.; Ward, E.W.B. Phytochemistry 1976, 15, 855-872.
22. Kuć, J.A. in "Encyclopedia of Plant Physiology"; Heitefuss, R.; Williams, P.H., Eds.; Springer: Berlin, 1976; Vol. 4, pp. 632-652.
23. Van Etten, H.D.; Pueppke, S.G. in "Biochemical Aspects of Plant-Parasite Relationships"; Friend, J.; Threlfall, D.R., Eds.; Academic: London, 1976; pp. 239-289.
24. Gross, D. Fortschr. Chem. Org. Naturst. 1977, 34, 187-247.
25. Kuan, T.L.; Erwin, D.C. Phytopathology 1980, 70, 333-338.
26. Keeling, B.L. Plant Disease 1982, 66, 334-335.
27. Keen, N.T.; Horsch, R. Phytopathology 1972, 62, 439-442.
28. Albersheim, P.; Valent, B.S. J. Cell Biol. 1978, 78, 627-643.
29. West, C.A. Naturwissenschaften 1981, 68, 447-457.
30. Ayers, A.; Ebel, J.; Finelli, F.; Berger, N.; Albersheim, P. Plant Physiol. 1976, 57, 751-759.
31. Ayers, A.; Ebel, J.; Valent, B.; Albersheim, P. Plant Physiol. 1976, 57, 760-765.
32. Ayers, A.; Valent, B.; Ebel, J.; Albersheim, P. Plant Physiol. 1976, 57, 766-774.
33. Ebel, J.; Ayers, A.; Albersheim, P. Plant Physiol. 1976, 57, 775-779.
34. Albersheim, P.; Darvill, A.G.; Sharp, J.K.; Davis, K.R.; Nothnagel, E.A.; Valent, B.S.; McNeil, M. "Complex Carbohydrates are Elicitors of Phytoalexin Accumulation"; Fourth International Congress of Plant Pathology; Melbourne, 1983, Abstract 243.
35. Ossowski, P.; Pilotti, A.; Garegg, P.J.; Lindberg, B. Angew. Chem. 1983, 95, 809-810.
36. Ziegler, E.; Pontzen, R. Physiol. Plant Pathol. 1982, 20, 321-331.

37. Wade, M.; Albersheim, P. Proc. Natl. Acad. Sci. USA 1979, 76, 4433-4437.
38. Desjardins, A.E.; Ross, L.M.; Spellman, M.W.; Darvill, A.G.; Albersheim, P. Plant Physiol. 1982, 69, 1046-1050.
39. Nichols, E.J.; Beckman, J.A.; Hadwiger, L.A. Plant Physiol. 1980, 66, 199-204.
40. Cline, K.; Albersheim, P. Plant Physiol. 1981, 68, 207-220.
41. Cline, K.; Albersheim, P. Plant Physiol. 1981, 68, 221-228.
42. Young, D.H.; Pegg, G.F. Plant Physiol. 1981, 19, 391-417.
43. Keen, N.T.; Yoshikawa, M. Plant Physiol. 1983, 71, 460-465.
44. Keen, N.T.; Yoshikawa, M.; Wang, M.C. Plant Physiol. 1983, 71, 466-471.
45. Davis, K.R.; Lyon, G.D.; Darvill, A.G.; Albersheim, P. Plant Physiol. 1982, 69, S-142.
46. Lyon, G.D.; Albersheim, P. Plant Physiol. 1982, 70, 406-409.
47. Hargreaves, J.A.; Bailey, J.A. Physiol. Plant Pathol. 1978, 13, 89-100.
48. Hahn, M.G.; Darvill, A.G.; Albersheim, P. Plant Physiol. 1981, 68, 1161-1169.
49. Nothnagel, E.A.; McNeil, M.; Albersheim, P.; Dell, A. Plant Physiol. 1983, 71, 916-926.
50. Bruce, R.J.; West, C.A. Plant Physiol. 1982, 69, 1181-1188.
51. Moesta, P.; Grisebach, H. Arch. Biochem. Biophys. 1981, 212 462-467.
52. Ebel, J.; Schmidt, W.E.; Loyal, R. Arch. Biochem. Biophys. submitted.
53. Moesta, P.; Grisebach, H. Arch. Biochem. Biophys. 1981, 211, 39-43.
54. Yoshikawa, M.; Yamauchi, K.; Masago, M. Physiol. Plant Pathol. 1979, 14, 157-169.
55. Dewick, P.M. in "The Flavonoids: Advances in Research"; Harborne, J.B.; Mabry, T.J., Eds.; Chapman and Hall: London, 1982; pp. 535-640.
56. Ebel, J.; Hahlbrock, K. in "The Flavonoids: Advances in Research"; Harborne, J.B.; Mabry, T.J., Eds.; Chapman and Hall: London, 1982; pp. 641-679.
57. Zähringer, U.; Ebel, J.; Mulheirn, L.J.; Lyne, R.L.; Grisebach, H. FEBS Letters 1979, 101, 90-92.
58. Leube, A.; Grisebach, H. Z. Naturforsch. 1983, 38 c, 730-735.
59. Zähringer, U.; Ebel, J.; Grisebach, H. Arch. Biochem. Biophys. 1978, 188, 450-455.
60. Ebel, J. in "Regulation of Secondary Product and Plant Hormone Metabolism"; Luckner, M.; Schreiber, K., Eds.; Pergamon: Oxford, 1979, pp. 155-162.
61. Börner, H.; Grisebach, H. Arch. Biochem. Biophys. 1982, 217, 65-71.
62. Hille, A.; Purwin, C.; Ebel, J. Plant Cell Reports 1982, 1, 123-127.
63. Ragg, H.; Kuhn, D.N.; Hahlbrock, K. J. Biol. Chem. 1981, 256, 10061-10065.
64. Lawton, M.A.; Dixon, R.A.; Hahlbrock, K.; Lamb, C.J. Eur. J. Biochem. 1983, 130, 131-139.
65. Yoshikawa, M.; Keen, N.T.; Wang, M. C. Plant Physiol. 1983, 73, 497-506.

66. Hahn, M.G.; Grisebach, H. Z. Naturforsch. 1983, 38 c, 578-582.
67. Keen, N.T. Physiol. Plant Pathol. 1971, 1, 265-275.
68. Yoshikawa, M.; Yamauchi, K.; Masago, H. Physiol. Plant Pathol. 1978, 12, 73-82.
69. Börner, H.; Schatz, G.; Grisebach, H. Physiol. Plant Pathol. 1983, 23, 145-152.
70. Stössel, P.; Lazarovits, G.; Ward, E.W.B. Can. J. Bot. 1980, 58, 2594-2601.
71. Ward, E.W.B.; Stössel, P.; Lazarovits, G. Phytopathology 1981, 71, 504-508.
72. Frank, J.A.; Paxton, J.D. Phytopathology 1970, 60, 315-318.
73. Beagle-Ristaino, J.E.; Rissler, J.F. Phytopathology 1983, 73, 590-595.
74. Lazarovits, G.; Stössel, P.; Ward, E.W.B. Phytopathology 1981, 71, 94-97.
75. Ward, E.W.B.; Lazarovits, G. Phytopathology 1982, 72, 826-830.
76. Grisebach, H.; Börner, H.; Moesta, P. Ber. Deutsch. Bot. Ges. 1982, 95, 619-642.
77. Moesta, P.; Hahn, M.G.; Grisebach, H. Plant Physiol. 1983, 73, 233-237.
78. Moesta, P.; Grisebach, H.; Ziegler, E. Eur. J. Cell Biol. 1983, 31, 167-169.
79. Moesta, P.; Seydel, U.; Lindner, B.; Grisebach, H. Z. Naturforsch. 1982, 37 c, 748-751.
80. Moesta, P.; Grisebach, H. Physiol. Plant Pathol. 1982, 21, 65-70.
81. Amrhein, N.; Gödeke, K.H. Plant Sci. Lett. 1977, 8, 313-317.
82. Keen, N.T.; Bruegger, B. in "Host Plant Resistance to Pests"; Hedin, P.A., Ed.; ACS Symposium Steries No. 62, American Chemical Society: Washington, D.C., 1977; pp. 1-26.
83. Cartwright, D.W.; Langcake, P.; Ride, J.P. Physiol. Plant Pathol. 1980, 17, 259-267.
84. Ward, E.W.B.; Lazarovits, G.; Stössel, P.; Barrie, S.D.; Unwin, C.H. Phytopathology 1980, 70, 738-740.
85. Grohmann, U.; Hoffmann, G.M. Journal of Plant Diseases and Protection 1982, 89, 435-446.
86. Stössel, P.; Lazarovits, G.; Ward, E.W.B. Phytopathology 1982, 72, 106-111.
87. Lazarovits, G.; Ward, E.W.B. Phytopathology 1982, 72, 1217-1221.

RECEIVED March 19, 1984

INDEXES

Author Index

Subject Index

A

E

F

G

Production by Anne Riesberg
Indexing by Deborah Corson
Jacket design by Anne G. Bigler
Elements typeset by Hot Type Ltd., Washington, D.C.
Printed and bound by Maple Press Co., York, Pa.

RECENT ACS BOOKS

"Advances in Pesticide Formulation Technology"
Edited by Herbert B. Scher
ACS SYMPOSIUM SERIES 254; 264 pp.; ISBN 0-8412-0840-9

"Structure/Performance Relationships in Surfactants"
Edited by Milton J. Rosen
ACS SYMPOSIUM SERIES 253; 356 pp.; ISBN 0-8412-0839-5

"Chemistry and Characterization of Coal Macerals"
Edited by Randall E. Winans and John C. Crelling
ACS SYMPOSIUM SERIES 252; 192 pp.; ISBN 0-8412-0838-7

"Conformationally Directed Drug Design:
Peptides and Nucleic Acids as Templates or Targets"
Edited by Julius A. Vida and Maxwell Gordon
ACS SYMPOSIUM SERIES 251; 288 pp.; ISBN 0-8412-0836-0

"Ultrahigh Resolution Chromatography"
Edited by S. Ahuja
ACS SYMPOSIUM SERIES 250; 240 pp.; ISBN 0-8412-0835-2

"Chemistry of Combustion Processes"
Edited by Thompson M. Sloane
ACS SYMPOSIUM SERIES 249; 286 pp.; ISBN 0-8412-0834-4

"Catalytic Materials:
Relationship Between Structure and Reactivity"
Edited by Thaddeus E. Whyte, Jr., Ralph A. Dalla Betta,
Eric G. Derouane, and R. T. K. Baker
ACS SYMPOSIUM SERIES 248; 470 pp.; ISBN 0-8412-0831-X

"NMR and Macromolecules:
Sequence, Dynamic, and Domain Structure"
Edited by James C. Randall
ACS SYMPOSIUM SERIES 247; 282 pp.; ISBN 0-8412-0829-8

"Geochemical Behavior of Disposed Radioactive Waste"
Edited by G. Scott Barney, James D. Navratil, and W. W. Schulz
ACS SYMPOSIUM SERIES 246; 413 pp.; ISBN 0-8412-0827-1

"Size Exclusion Chromatography: Methodology and
Characterization of Polymers and Related Materials"
Edited by Theodore Provder
ACS SYMPOSIUM SERIES 245; 392 pp.; ISBN 0-8412-0826-3

"The Chemistry of Solid Wood"
Edited by Roger M. Rowell
ADVANCES IN CHEMISTRY SERIES 207; 616 pp.; ISBN 0-8412-0796-8

"Polymer Blends and Composites in Multiphase Systems"
Edited by C. D. Han
ADVANCES IN CHEMISTRY SERIES 206; 400 pp.; ISBN 0-8412-0783-6